SCIENTIFIC GERMAN

by GEORGE E. CONDOYANNIS

Professor of Modern Languages
Saint Peter's College

SCIENTIFIC GERMAN

A CONCISE DESCRIPTION OF THE STRUCTURAL ELEMENTS
OF SCIENTIFIC AND TECHNICAL GERMAN

ROBERT E. KRIEGER PUBLISHING COMPANY
HUNTINGTON, NEW YORK
1978

Original Edition 1957
Seventh Printing 1966
Reprint 1978, with corrections and additions

Printed and Published by
**ROBERT E. KRIEGER PUBLISHING CO., INC.
645 NEW YORK AVENUE
HUNTINGTON, NEW YORK 11743**

Copyright © 1957 by
JOHN WILEY & SONS, INC.
Reprinted by Arrangement

All rights reserved. No reproduction in any form of this book, in whole or in part (except for brief quotation in critical articles or reviews), may be made without written authorization from the publisher.

Printed in the United States of America

Library of Congress Cataloging in Publication Data

Condoyannis, George Edward, 1915-
  Scientific German.

  Reprint of the 1966 issue of the edition first published in 1957 by Wiley, New York.
  1. German language—Technical German. 2. German language—Grammar—1950-   I. Title.
[PF3120.S3C6 1978]   438'.2'421   77-16570
ISBN 0-88275-644-3

Preface

This book is intended to fill a need not ordinarily met by elementary language texts. The young scientist or engineer, as well as the student preparing for these professions, is frequently faced with the problem of acquiring rapidly and directly a reading knowledge of German adequate to cope with technical articles and books in his field. This textbook is designed to be used from the very beginning in conjunction with such unsimplified scientific and technical reading matter and with a dictionary. It is meant primarily for beginners with no previous knowledge of German, but can also be used by those who wish to refresh their knowledge of the language. It can be used in classes or for individual study.

The student at first tends to think of his difficulties in terms of individual words. These can be looked up with relative ease, assuming a certain amount of language sense, in a dictionary. But there are other difficulties for which the dictionary offers no solution. In fact, it is possible to know the exact meaning and even the precise translation of each word and still be unable to combine these elements in any way that will convey the sense of the original. To cope with this type of difficulty, the student must master the basic structure of the language. This is presented here, in a form specifically tailored to the reading aim.

A reader is primarily interested in understanding a form that occurs in a passage before him, whereas the traditional language textbook shows him how to reproduce this form for use in speaking or writing. Thus the aim of this book is to answer the question: "How do I know whether a form used by an author is the present or the past tense of the verb?" rather than the more traditional question: "How do I form the past from a present form that I have already learned?" In short, the aim here is analytic rather than synthetic.

A perennial difficulty in beginners' language textbooks is that of grammatical terminology. No discussion of the structure of a language is possible without it, yet the layman is rarely familiar with it. Hence all grammatical terms are explained when they are introduced, and the number is kept to a minimum. At the same time every attempt has been made to use more or less self-explanatory terms: stem-changing verbs, endingless adjectives, variable and constant verb forms, subjectless passive, etc.

Years of classroom experience have shown that many students have only vague notions regarding some of the more sophisticated grammatical concepts, notably the passive voice and the impersonal verb. Mere mechanical handling of such difficulties has proved to be far from satisfactory and gives the student no sense of security. This book, therefore, attempts to impart a real, basic understanding of such difficulties.

It was also found that a reading-aim textbook must cope more seriously than the ordinary beginner's book with the problem of what to take up first, since almost any construction may occur in the very first sentence of the first technical article the student tries to read. The solution has been to start with a bird's eye view of the structure of the German language. This is followed by a lesson on basic elements: the grammatical units and sentence patterns encountered again and again in technical prose. The subsequent lessons serve mainly to fill in the details of the over-all pattern. In fact, the chapters of the book are not lessons in the ordinary sense. Toward the end they take on much of the character of a reference grammar to be consulted as the need arises.

It may not be obvious that reading in a foreign language is an acquired skill of the highest order. It takes extensive practice. This book and the dictionary are the tools for the job but, to achieve any real facility, the only sure course is to *read:* a hundred, perhaps two hundred pages of material on various aspects of your own field, preferably material written by different authors. But the skill, once acquired, must also be maintained by using it, year after year, and at the same time constantly improving it.

ACKNOWLEDGMENTS

To a great extent this book is the outgrowth of many years of experience teaching reading-aim courses in both German and French at the Massachusetts Institute of Technology, side by side with my colleague Professor William N. Locke, who encouraged me to undertake the writing of this textbook. Particular thanks are also due the Institute's Department of Modern Languages for aid in preparing the manuscript. My gratitude is also extended to the various publishers who graciously gave permission to use portions of their technical articles and books for the reading selections.

GEORGE E. CONDOYANNIS

Jersey City, N. J.
March, 1957

Contents

Lesson

1	Pronunciation and Spelling	1
2	General Remarks on Structure	11
3	Basic Elements	20
4	The Noun-Adjective System	30
5	Pronouns	47
6	Word Order—Part One: Normal; Inverted; Dependent	57
7	Word Order—Part Two: Relative Clauses; Special Difficulties	67
8	Verbs: A. Simple Tenses	74
	B. Compound Tenses	81
9	Verbs: A. Reflexive	85
	B. Passive	87
	C. Impersonal	90
10	Verbs with Prefixes	95

ix

11	Constructions Involving Infinitives	105
	A. With "zu"	105
	B. Without "zu"	107
	C. Modal Auxiliaries	108
12	Adjectives and Adverbs; the Extended Modifier Construction	115
13	Prepositions	129

Appendix

1	Details Regarding Demonstratives	136
2	Prepositions and Their Cases	140
3	The Alphabet	143
4	Numbers	145
5	Most Frequent Words	147
6	Principal Parts of Simple Stem-Changing Verbs	151
7	Alphabetical List of Verb Forms	157

LESSON **1**

Pronunciation and Spelling

There are many similarities in vocabulary and grammatical structure between German and English, but German pronunciation differs markedly from English. German pronunciation is usually very precise, and individual words are relatively easy to identify as separate units, since they are not run into each other. Yet special study is needed to imitate and understand spoken German, and this involves extensive modifications of English speaking habits.

In the following tables the sounds represented by German letters are illustrated by the nearest American equivalent. The resulting pronunciation will not necessarily be understandable to a German, but may suffice for communication of German words between Americans.

TYPE AND SCRIPT

Nearly all scientific and technical publications—and newspapers— now use Roman type exactly like that used in English. Other books and many newspapers were formerly printed in the so-called German or Gothic or Old English type, in which the same letters are given a very ornate form, as in the title of many American newspapers (e.g., The New York Times). For the reader who may be interested, this alphabet is tabulated in Appendix 3. As for handwriting, Germans use the same letters as we do. A special script corresponding to the ornate German type was once current, but is no longer in use.

TABLE 1. LETTERS ALWAYS REPRESENTING SIMILAR SOUNDS IN ENGLISH AND GERMAN

(Illustrated by Words of Similar Sound)

LETTER	ENGLISH	GERMAN	MEANING OF GERMAN
f	fell	Fell	(hide, skin)
h (pronounced)[1]	house	Haus	(house)
h (silent)[1]	ah	ah	(ah)
k	kit	Kitt	(cement)
l (as in English mi*ll*ion)	million	Million	(million)
m	man	Mann	(man)
n	note	Not	(distress)
p	pact	Pakt	(pact)
ph	photograph	Photograph	(photographer)
s (hissing sound)	glass	Glas	(glass)
s (buzzing sound)	miserable	miserabel	(miserable)
t	tote	tot	(dead)
x[2]	taxi	Taxi	(taxi)

[1] h is silent *only* where—as often in English—it is used to indicate that a preceding vowel is "long" or sustained: ah, oh, etc.

[2] This includes x at the beginning of a word, e.g., *Xylose*.

Table 2. Letters Sometimes Representing Similar Sounds in English and German

(See Other Uses of These Letters in Table 3)

LETTER

b (except in final position)	bank	Bank	(bank; bench)
c (like k before a, o, u in words of Latin origin)	calcium	Calcium	(calcium)
c (in a few words of French origin)	balance	Balance	(balance)
ch (like k before a, o, u, l, r in words of Greek origin)	character chrome	Charakter Chrom	(character) (chromium)
ch (like sh in words of French origin)	chef	Chef	(head, boss)
d (except in final position)	odor	oder	(or)
g (as in go, except in final position)	gift	Gift	(poison)
ng (as in singer, never as in finger)	sang	sang	(sang)
v (in words of Latin origin)	November	November	(November)

Table 3. Letters Representing Different Sounds in English and German

LETTER	ENGLISH	GERMAN	MEANING OF GERMAN
b (in final position)[1]	p	ab	(off, away)
c (before e, i, y, ä)	ts, tz	Centrum	(center)
ch (after a, o, u)	No equivalent. A roughened *h* sound produced at the back of the mouth.	Nacht Loch Buch	(night) (hole) (book)
ch (after e, i, y, ä, ö, ü and consonants; initially before e, i, ä)	No equivalent. A roughened *h* sound produced in the same position as the *h* of the word *human*.	Blech Licht Milch durch	(sheet metal) (light) (milk) (through)

---

[1] Independent elements ending in b, d, g, s retain the final-position pronunciation even when they are used as components of longer words: Verabredung (b like p), Gasentladung (s like ss).

TABLE 3 (Continued)

LETTER		ENGLISH	GERMAN	MEANING OF GERMAN
chs		x	wachsen	(grow)
d	(in final position)[1]	t	Rad	(wheel)
g	(in final position)[1]	k	Tag	(day)
g	(in the ending **-ig**)	same as ch after e, i, etc.	staubig	(dusty)
g	(in a few French words)	z of a*z*ure	Etage	(stage, level)
gn		gn of A*gn*es	Gnade	(mercy)
j		y of *y*es	ja	(yes)
kn		cn of a*cn*e	Knoten	(knot)
pf		pf of to*pf*light	Pfund	(pound)
pn		pn of to*pn*otch	pneumatisch	(pneumatic)
ps		ps of ta*ps*	psychologisch	(psychological)
qu		kv	Quelle	(source)
r		No equivalent. Sound can be made by trilling the tip of the tongue.	rot wer hart Wasser	(red) (who) (hard) (water)
s	(initially, followed by a vowel)	z	so	(so)
s	(in final position)[1]	s of ga*s*	des Tisches	(of the table)
sch		sh	schnell	(quick)
sp	(beginning a stressed or root syllable)	shp as in fle*shp*ot	sprechen	(speak)
st	(beginning a stressed or root syllable)	sht as in A*sht*abula	Stein	(stone)
th[2]		t	Thema	(theme)

[1] Independent elements ending in b, d, g, s retain the final-position pronunciation even when they are used as components of longer words: Ver*a*bredung (b like p), Gas*e*ntladung (s like ss).

[2] Never like *th* in English, a sound which does not exist in German.

Table 3 (Continued)

LETTER	ENGLISH	GERMAN	MEANING OF GERMAN
v (except in some foreign words)	f	vor	(before)
w	v	wo	(where)
z, tz	ts of ca*ts*	zehn	(ten)
		Katze	(cat)

THE VOWELS

German vowels differ from English vowels in several important respects.

1. Each of the symbols a, e, i, o, u stands for one and only one sound.
2. Each of these symbols may indicate a long or a short pronunciation of the sound without affecting the quality of the sound itself.
3. The sounds represented by the vowel symbols are pure and must retain exactly the same quality regardless of how long they are drawn out in pronunciation; there must be no off-glides to other sounds (as there always are in English).

Indicating Vowel Length

Generally speaking, a vowel followed by a single consonant *is long* (especially if this consonant is followed by another vowel) and a vowel followed by two or more consonants is short:

>   Gas, Gase (long **a**)   Gasse (short **a**)
>   (gas, gases)   (alley)

But this principle cannot be consistently applied in all situations and additional devices are used to indicate *length:*

Doubling (used for **aa, ee** and **oo** only)
Silent **h** (**ah, eh, ih, oh, uh, äh,** etc.)

In the case of **i**, a special combination **ie** is also used and is equivalent to **ih** (**Nieren** rhymes with **ihren**).

Table 4. The Vowel Sounds

(The English equivalents of vowels are particularly hard to define because of the great variations in vowel pronunciation within the United States, not to mention British English. Careful imitation of an actual speaker or of special language records, with correction and explanation by a trained language teacher, is the only way toward correct pronunciation.)

LETTER	APPROXIMATE ENGLISH EQUIVALENT	SAMPLE WORD	MEANING OF GERMAN
a (*long*)	a in f*a*ther	Staat	(state)
a (*short*)	The same sound cut short, approx. like o in r*o*ck	Stadt	(city)
e (*long*)	a in m*a*y (*without the final off-glide to* y)	Beet	(flower bed)
e (*short*)	e in b*e*t	Bett	(bed)
e (unstressed, esp. in final position)[1]	a in sof*a*	Klasse	(class)
i (long)	ee in s*ee*n	Lid; Lied	(eyelid; song)
i (short)	i in un*i*que	litt	(suffered)
o (long)	o in gl*o*w (*without the final off-glide to* w)	Sohlen	(soles)
o (short)	au in t*au*t	sollen	(are supposed to)
u (long)	oo in m*oo*d	Ruhm	(fame)
u (short)	oo in m*oo*t	dumm	(stupid)

y (*in words of Greek origin like* ü: **Analyse**, *otherwise like German* **i**; *both pronunciations can occur as either long or short sounds*).

---

[1] Final e *must* be pronounced. It may make an important difference in meaning: sagt (says), sagte (said), ging (went), ginge (would go); der Zweck (purpose), die Zwecke (tack).

TABLE 5. VOWEL COMBINATIONS (DIPHTHONGS)

ai, ei, ay, ey	i in b*i*te	Saite	(musical string)
		Seite	(side)
		Bayern	(Bavaria)
		Meyer	(*family name*)
au	ou of h*ou*se	Haus	(house)
eu	oi of expl*oi*t	heute	(today)

VOWEL MODIFICATION (UMLAUT)

ä is equivalent to **e** in any situation:
    fällt rhymes with **schellt** (short vowel)
    Zähne rhymes with **Lehne** (long vowel)
äu is equivalent to **eu,** so that:
    Häute (plural of **Haut,** skin) sounds exactly like **heute** (today).
ö is a modification of the sound **o** toward the sound **e** (long or short).
    To pronounce it, get set to pronounce German **e** and pout your lips.
ü is a modification of the sound **u** toward the sound **i** (long or short).
    To pronounce it, get set to pronounce German **i** and pout your lips.
Warning. Umlaut (i.e., the modification of **a** to **ä**, etc.) is of basic importance in reading and must not be overlooked. It may make the difference between two words (**zahlen,** to pay vs. **zählen,** to count) (the dictionary lists them in this order), or it may indicate common grammatical variations, such as

Ofen	oven	→ **Öfen**	ovens	(singular to plural)
alt	old	→ **älter**	older, elder	(comparative of an adjective)
fallen	to fall	→ **fällt**	falls	(verb tense form)
wußte	knew	→ **wüßte**	would know	(different verb tense forms)

(*See also* Appendix 3.)

ONE SPECIAL CONSONANT SYMBOL

The symbol ß is used under certain conditions as the equivalent of **ss: wußte = wusste.**

STRESS AND ACCENTUATION

German, like English, stresses one syllable in a word and tends to understress the rest (English, therMOMeter; German, ThermoMEter). But the *un*stressed syllables are not slurred over as much as in English and the vowels in them must retain their original quality, i.e., every *a, o, i* or *u* in an unstressed syllable still sounds like German **a, o, i,** or **u: separat, Definition,** and most *e*'s retain their original sound, unless they are final or in the ending *-er*, in which case they sound like the *a* of sof*a*.

Most German words are stressed on the "root syllable," i.e., on the syllable resembling the one-syllable word to which various deriving elements have been added. The root "Satz," with its alternate "setz," is stressed in the words:

  S**ä**tze   Ges**e**tz
  S**a**tzung   ges**e**tzlich
  Bes**a**tzung Ges**e**tzlichkeit
  Ers**a**tz   Vers**e**tzungen
         Vers**e**tzungsfähigkeit, etc.

But if an element that can exist as a separate word is put ahead of the root syllable, then this first element is usually stressed, on its own root syllable:

  Auf/satz   **Ein**satz/bereitschaft
  durch/gesetzt ent**gegen**/gesetzt
  zus**am**men/zu/setzen

The same principle applies essentially to compound words: the first component receives the main stress (on its own root syllable) and the other components then have a secondary stress on their respective root syllables:

G**eschlechts**/*le*ben sex life
Lebens/ver*sich*erungs/ge*sell*schaft life insurance company

This is substantially what we do in English if we consider as a unit such combinations as "life insurance company."

Words taken into German from foreign languages, especially those of Latin origin, often closely resemble English words but are almost always stressed differently, generally on the last syllable:

GenerRAL, negaTIV, BioloGIE,* MathemaTIK.

EXERCISES

It is suggested that you write out a translation into good technical English of all the exercises in this book. Use a dictionary. In early lessons, not all of the difficulties you encounter will have been explained. This is inevitable, since the material used for the exercises is taken from technical journals and similar sources—as indicated—virtually unsimplified. For this reason a few explanatory notes are included with the first few exercises.

EXERCISE
PHYSIK

Einführung

Physikalische Vorgänge sind:

1. Bewegungsvorgänge.
2. Wärmevorgänge.
3. Schall.
4. Licht.
5. Magnetismus.
6. Elektrizität.
7. Gravitation = Erscheinungen der Schwere.

Bei[1] den physikalischen Vorgängen verändert[2] sich[2] nur der Zustand des Körpers, während bei den chemischen Vorgängen sich[2] der Stoff des Körpers verändert.[2]

---

* Final **ie** is pronounced like German long **i** (hence like English *ee*) in most foreign words, especially if they came into German by way of French. Those considered to be taken directly from Latin are stressed on the syllable *before* the **ie**, while the **ie** itself is, by way of exception, split into two syllables, sounding, roughly, like the final *ia* of the word California: MaTEri e, LIni e.

Allgemeine Eigenschaften der Körper

Undurchdringlichkeit, Raumausfüllung, Stofflichkeit, Teilbarkeit, Gewicht, Aggregatzustand (fest, flüssig, gasförmig).

Moleküle sind die kleinsten Teilchen eines Stoffes, die[3] noch die gleichen Eigenschaften haben wie[4] der Stoff selbst.

Moleküle lassen[5] sich[5] chemisch in Atome (wahrhaft unteilbare Grundstoffe) zerlegen.[5]

Kohäsion = Zusammenhangskraft der Moleküle eines und desselben Körpers.

Sie ist bei den einzelnen Körpern verschieden.

Feste Körper: große Kohäsion, selbständige Gestalt, bestimmtes Volumen.

Flüssige Körper: geringe Kohäsion, keine selbständige Gestalt, bestimmtes Volumen.

Gasförmige Körper: keine Kohäsion, sondern[6] Expansion, keine selbständige Gestalt, kein bestimmtes Volumen.

> Taken from *Physik* bearb. v. Walter Baack, Ing., Oberlehrer an der höheren Marinefachschule für Gewerbe und Technik, Wilhelmshaven, Verlag Moritz Diesterweg, Frankfurt a. M., 1931, p. 5.

Notes for Exercise

[1] **Bei,** in.
[2] **Verändert sich** reflexive verb. Look under **verändern** *v.r.*
[3] **Die** here means "which."
[4] **Wie** here means "as."
[5] **Lassen sich . . . zerlegen** can be decomposed.
[6] **Sondern,** but on the contrary.

LESSON 2

General Remarks on Structure

There are three large divisions of structure: the noun-adjective system, the verb system, and word order.

A. THE NOUN-ADJECTIVE SYSTEM

German nouns belong to one of three classes or *genders;* they also have a *plural* form. Both the singular and the plural form vary for case, i.e., they have changes in form for reasons similar to the changes from *he* (subject form) to *him* (object form) or *his* (possessive form).

The adjective in German always precedes the noun it modifies and agrees with it in *gender, number* (singular or plural), and *case* (subject or object form, etc.), i.e., it varies in form corresponding to the variations of the noun it modifies. There is usually more variation in the adjective than in the noun, and in any adjective-noun combination, it is the adjective that shows, by its endings, the case, number, and gender of the noun.

Pronouns also vary for gender, number, and case, often just as drastically as in English *I—me, we—us.*

B. THE VERB SYSTEM

German verbs resemble English verbs in that they fall into two large classes according to the formation of their past tense: *write-wrote* (internal change; we shall call them "stem-changing" verbs) vs. *talk-talked* (addition of a suffix or an "ending"). German, like English, has some *simple* (i.e., one-word) *tenses*, such as "present" (write, writes) and "past" (wrote), and a number of combinations or *compound tenses* made up of an *auxiliary verb* and special forms of the verb in question:

> I have written they were writing
> he will write it will be written

German has fewer such combinations than English; most German compound tenses perform the functions of two or more English tenses. On the other hand, the German verbs have more variation for *person*, i.e., more different forms to correspond to different grammatical subjects (I, you, he, we, they, etc.). Thus, while English has only two variations within the simple present tense:

> I, you, we, they *write* he, she, it *writes*

German has three variations, because the form to go with the subject "I" is different from the one that goes with the subjects "you," "we" and "they."

Note. In fact, it also has two additional forms to go with special second-person pronouns (one singular, one plural) used in speaking familiarly or intimately. Since these do not occur in scientific and technical writing, they are left out of consideration throughout this book.

Like their English counterparts, German verbs are best learned by memorizing their principal parts, such as sing, sang, sung; write, wrote, written; go, went, gone, etc. The first form in each of these sequences is the *infinitive* (the form listed in the dictionary), the second is the *past tense*,* and the third is the *past*

---

* Also called the *imperfect tense* in many textbooks.

participle. The uses of these forms are, with slight modifications, substantially the same as in English.

singen, sang, gesungen **schreiben, schrieb, geschrieben**
gehen, ging, gegangen

C. WORD ORDER

It is in word order that German differs most radically from English.

In most respects German word order is rigidly fixed. For example, if there are expressions of *time* and *place* in one clause, they come in that order, and other adverbial modifiers come between them.

Ich fahre	**jetzt**	**mit dem Autobus**	**nach Hause.**
I am going	now	by bus	home.

Direct and indirect objects are usually found in the same relative position as in English, if we ignore the preposition *to* with the indirect:

Ich	**gab**	**dem**	**Mann**	(or **gab ihm**)	**die Auskunft.**
I	gave	the	man	(*or* gave him)	the information.
Ich	**gab**	**sie**	**dem Mann**	(or **sie ihm**).	
I	gave	it	to the man	(*or* it to him).	

For the most part, however, German word order is a question of where to find the verb. There are two kinds of verb forms: *variable* (changing according to the subject: I *have,* he *has*) and *constant* (never changing: I have *gone,* he has *gone*). In German the variable is either in first position (questions, commands, "if-less if-clauses"), in second position (ordinary statements in independent clauses), or in last position (dependent clauses.) Moreover, in a dependent clause, all verb forms are clustered at the end and their order is precisely the reverse of what it would be in English.

First position: **Hat** man etwas Neues gefunden?
Has one found something new?

Second position: Man **hat** etwas Neues gefunden.
One *has* found something new.

Last position: . . . , weil man etwas Neues gefunden **hat**.
because one *has* found something new.

In one respect German word order is more flexible than English: any sentence element, be it subject, adverbial expression, or object, can be put first for the sake of emphasis; the verb is then second. Thus, if the subject does not begin the sentence, it comes *after* the verb. This means that from the English-speaking reader's point of view, and especially from the translator's, the verb seems to come either too soon (before the subject) or too late (at the end).

The beginning reader must, therefore, get used to these basic patterns:

1. *Subject   Variable Verb   Pronoun Objects* (if any)
Adverbs (time-manner-place)   *Noun Object* (if any)   *Constant Verb Forms* (if any)

Example

Die Körper   **können**   **sich**
(subject)   (variable verb)   (object pronoun)
im Zustand der Ruhe oder Bewegung   **befinden.**
(adverbial phrase)   (constant verb)

Bodies can be (*literally,* can find themselves) in a state of rest or of motion.

2. *Adverb* or *Object   Variable Verb* (pronoun objects and adverbial expressions are often inserted here)   *Subject Other elements* as above

Cf. English: (Adverb)   (Verb)   (Subject)
             Only then   was   it   possible. . . .

Example

Allgemein   **wird**   **die Lehre vom Gleichgewicht**
(adverb)   (var. verb)   (subject with modifying
                                     phrase)

Word Order

mit Statik	bezeichnet.
(adverbial phrase)	(constant verb)

Generally the study of equilibrium is designated as statics.

3. *Variable Verb   Subject   Other elements,   so Variable Verb   Subject   Other elements* (This is an "if-less if-clause": start your translation with "if")

Cf. English: Were this the case, then. . . . Equivalent to: *If* this were the case, then . . .

Example

Beträgt	**die Oberfläche eines Körpers**	F cm²,
(variable verb)	(subject with modifier)	(object)
so kann	daraus	
(variable verb)	(adverb)	
die Größe der Kraft	**ermittelt werden.**	
(subject, with modifier)	(constant verbs)	

If the surface of a body amounts to F sq cm, then the magnitude of the force can be determined from that.

4. Dependent Clause: *Conjunction   Subject   Other elements   Constant Verb Forms   Variable Verb*

Example

Wenn	**ein**	**Maschinenteil**	**sich**	**bewegt** . . .
(conjunction)		(subject)	(object pronoun)	(variable verb)
If	a	part of a machine		moves . . .
. . . **die Kraft,**	**welche**	**auf die Oberfläche**	**wirkt.**	
	(relative pronoun subject)	(adverbial phrase)	(variable verb)	
. . . the force	that	acts on the surface.		

Note. All dependent (or subordinate) clauses are set off by commas in German. This is of great help to the reader, especially in seeking out verb forms that come in last position.

For details on word order, see Lessons 6 and 7.

D. TRANSLATION

One can learn to read German and understand it without translating it. Translation is a separate step and an additional technique. It is impossible, in any language, to rely on a word-for-word correspondence, since we translate ideas and not words. The translator must not only thoroughly understand the German but also have a complete command of English. Generally speaking, a close translation is probably preferable to a "free" one, as long as it does no violence to the English language. In any case, the translator *must always know* the literal meaning of a German expression before he can rephrase the idea into a "free" translation.

E. USE OF THE DICTIONARY

A dictionary cannot be used intelligently without a knowledge of the structure of the language—and of English. For example, most dictionaries list the irregular past tenses of verbs, but they do not list these again with prefixes (*told* but not *retold, went* but not *underwent,* etc.). This means that the reader, in German also, must be able to detach prefixes, look up the verb form to determine its origin (the dictionary will only refer you to the main entry), reattach the prefix and look up the main entry (in English, for example, to find the meaning of *underwent:* look up *went,* trace it to *go,* reattach *under,* and look up *undergo.* This is the procedure to be followed with many German verb forms also).

Dictionaries designate each use of a verb as *transitive* (abbreviated *v.t.* = used with a direct object), *intransitive* (abbreviated *v.i.* = used without an object) and *reflexive* (*v.r.* = used with a reflexive pronoun: in German usually *sich*). Since the translation and meaning of the same German verb may differ according to whether it is used transitively, intransitively, or reflexively, it is important to determine its use in the text before looking it up.

Dictionaries list the singular of nouns, but not their plural; they list adjectives without their endings and in German this is also the corresponding adverb; they list the positive but not the comparative or superlative degree (*large* but not *larger* and *largest*). Thus in many instances the reader must be able to reduce a form he finds in a text to its dictionary form before he can look it up.

It is a rare word that has only one meaning, and it is an even rarer word that has exactly the same range of meanings as a corresponding word in another language. Dictionaries list as many meanings, and their translations, as they can do in conformity with space limitations, but they cannot foresee all the contextual difficulties that may come up in a given phrase or sentence. Once the reader has an idea of the meaning of the word, he must still fit this meaning into the context in its right form (singular or plural, past or present, etc.) and in such a way as to achieve a smooth English wording.

EXERCISE
A. MECHANIK

Einleitung

Die Mechanik ist die Lehre von den Kräften und von den Bewegungen der Körper.

Kraft ist die Ursache einer Bewegung oder Bewegungsänderung. Die Kräfte können bewegend oder hemmend sein (z.B.[1] Schwerkraft bzw.[2] Reibung).

Die Körper können sich[3] im Zustand der Ruhe (Gleichgewicht) oder der Bewegung befinden.[3] Die Körper selbst können **fest, flüssig** oder **gasförmig** sein.

Allgemein wird die Lehre vom Gleichgewicht mit **Statik**, die Lehre von der Bewegung mit **Dynamik** bezeichnet.[4]
Dynamik Fester Körper.—Bewegungslehre
Bewegung ist die Ortsveränderung eines Körpers. Man[5] unterscheidet:
1. gleichförmige Bewegung;
2. ungleichförmige Bewegung,
   (a) gleichförmig-beschleunigte Bewegung,
   (b) gleichförmig-verzögerte Bewegung.

Nach der Form des zurückgelegten Weges unterscheidet man:
1. geradlinige Bewegung;
2. krummlinige Bewegung (Kreisbewegung).

Adapted from Hans Jönck, *Lehrhefte für Mechanik und Festigkeitslehre*, Höhere Marinefachschule für Gewerbe und Technik, Kiel, 1929.

B. KRAFT UND DRUCK

Wenn ein Maschinenteil sich bewegt,[6] dann ist dafür eine Ursache vorhanden, die[7] wir **Kraft** nennen. Zur Angabe einer Kraft brauchen wir: **Angriffspunkt, Richtung** und **Größe**. Die Größe wird in der Technik in kg (Kilogramm) gemessen.

Oft wirken eine große Anzahl gleichgerichteter Kräfte senkrecht auf die Oberfläche eines Körpers. In diesem Falle spricht man von **Druck** und versteht darunter[8] die **Kraft, die**[7] **auf eine Flächeneinheit ausgeübt wird**. Der Druck erhält[9] die Bezeichnung p; man mißt[10] ihn in der Technik meistens in kg/cm^2.

Beträgt die Oberfläche eines Körpers F cm^2 und der Druck, der auf ihr lastet, p kg/cm^2, so kann daraus die Größe der Kraft ermittelt werden, welche auf die Oberfläche des Körpers wirkt.

Die Größe der Kraft ist das Ergebnis: Druck mal Fläche $P = p \cdot F$ kg.

Adapted from W. Seidel, *Maschinenkunde——Leitfaden für die Maschinistenmaatenlehrgänge an den Marineschulen Kiel und Wesermünde*, 1937.

Exercise 2

Notes for Exercise

[1] **Z. B., zum Beispiel,** for example, *e.g.*

[2] **Bzw., beziehungsweise,** or as the case may be, respectively: gravity or friction, respectively; gravity, or in the other case, friction.

[3] **Sich befinden,** reflexive verb, to be (located, found).

[4] The past participle **bezeichnet** is understood after **Statik** also. Generally speaking, if a word is to be used twice and is left understood one of the two times, German will leave it understood in its first occurrence and mention it in the second.

[5] **Man unterscheidet,** We distinguish, one distinguishes. See Lesson 4 E.

[6] **Sich bewegen,** reflexive verb, to move, to be moving. Many German reflexive verbs correspond to English intransitive verbs.

[7] **Die** is a "relative pronoun," introducing a dependent clause: "which we call force," "which is exerted. . . ." Forms like **der, die, das, dem,** and **den** following a comma and starting a clause with verb(s) in *last* position are usually relative pronouns.

[8] . . . **Man** . . . **versteht darunter** (idiom), we mean by it.

[9] **Erhalten** normally means to receive, get: The pressure gets (*or* is given) the designation p.

[10] **Mißt.** The present tense of some stem-changing verbs involves an internal change in the third person singular. Such forms are listed in the dictionaries, with a reference to the infinitive.

LESSON 3

Basic Elements

A. RECOGNITION OF NOUNS

All German nouns (i.e., names of persons, places, things) are *capitalized*. Moreover, all other parts of speech (verbs, adjectives) used as nouns must be capitalized. Thus there can never be any doubt as to whether a word is a noun or not, and the beginner must learn to rely constantly on this principle:

das Haus	the house	**arm** (adj.) poor: **die Armen** the poor
die Säure	the acid	**unternehmen** (verb) to undertake
der Punkt	the point	**das Unternehmen** the undertaking, enterprise

B. GENDER AND ARTICLES

All German nouns are divided into three classes, or "genders."
A "masculine" noun is one preceded by the article **der**.
A "feminine" noun is one preceded by the article **die**.
A "neuter" noun is one preceded by the article **das**.
These are the forms of the definite article "the" in German and serve as a convenient gender identification tag. (In the

subject-form, i.e., nominative case. See Lesson 4 for other forms.) Thus it is advisable always to learn every German noun with its article, hence not **Anfang**, but **der Anfang** (beginning), not **Mitte** but **die Mitte** (middle) and not **Ende**, but **das Ende** (end). This is, in the last analysis, the only foolproof way of remembering the gender of a noun.

It is important to know the gender of a noun because this often helps to identify whether it is a singular or plural form and because many modifiers of nouns and many words in other parts of the sentence referring to a particular noun will have the same gender as the noun.

C. BASIC AUXILIARY VERB FORMS

In English the verb "to be" when referring to present time has three forms, depending on the subject. German verbs likewise have three such forms:

		English		German	
		Subject	Verb	Subject	Verb
Singular	first person	I	am	ich	**bin**
	third person	he		er*	
		she		sie	
		it	is	es	**ist**
		any singular noun		(noun)	
Plural	first person	we		wir	
	second person	you		Sie	
	third person	they	are	sie	**sind**
		any plural noun		(noun)	

Two other basically important verbs are:

haben	*to have*	**werden**	*to become*
ich **habe**	wir	ich **werde**	wir
	Sie **haben**		Sie **werden**
er **hat***	sie	er **wird***	sie

* For convenience in tabulation, the third person singular of verbs is usually indicated as, for example, **er ist**; it is to be understood that the verb form that accompanies **er** also accompanies the subjects **sie** and **es**.

The verbs **sein, haben** und **werden,** besides being independent verbs with the meanings given, are also tense-forming auxiliaries, that is, they are used in combination with constant (i.e., invariable) forms of other verbs to make up new tenses (*cf.* English he *has* seen, it *is* said, etc.)

The verb **sein** (like its English equivalent "to be") is highly irregular. The other two verbs given here, **haben** and **werden,** are more typical in their forms and permit these general observations:

To distinguish between **sie** (she, it) and **sie** (they), we must look at the corresponding verb:

    sie **hat**   she, it *has*   sie **haben**   they *have*

The "general form" of the verb, the one by which we refer to it and the one listed in the dictionary, is called the infinitive. English "to be," German **sein.**

For all verbs except **sein** the plural form of the present tense is identical with the infinitive.

The verb **haben** illustrates the typical endings for 3^d person singular (**-t: hat**) and plural (**-en: haben**), which, because of their enormous frequency, are of basic importance to the reader.

The subject pronouns **er** and **sie** mean *he* and *she* when referring to persons; otherwise their English equivalent is *it:*

Masculine Nominative. Wo ist **der** Bleistift? Hier ist **er.**
                           Where is the pencil? Here *it* is.

Feminine Nominative. Wo ist **die** Feder? Hier ist **sie.**
                           Where is the pen? Here *it* is.

D. CONSTANT VERB FORMS

There are two non-varying or constant verb forms, which have close parallels in English:

1. The *infinitive,* which in German is characterized by the ending **-en** (sometimes **-n**).

gehen (to) go   **treiben** (to) drive   **verwandeln** (to) transform

2. The *past participle* (equivalent to English forms like

broken, spoken, ridden, driven, written, etc.) Most German past participles are characterized by the prefix **ge-**:

gezeigt shown   **getrieben** driven   **ausgetrieben** driven out

But past participles of verbs not stressed on the first syllable do not have this **ge-**. This includes:

(*a*) verbs with unstressed prefixes like **be-, ent-, emp-, er-, ver-, zer-: betrieben, verwandelt**

(*b*) verbs ending in **-ieren: konzentriert** (This includes an enormous number of verbs based on Latin roots and easily identified by the reader: **studieren, kontrollieren, reduzieren, analysieren, desinfizieren, identifizieren**. The accent is always on the **ie** before the **r**: kon-trol-LIE-ren.)

(*a*). *Compound Tenses*

With **haben**:

The present tense forms of **haben** combined with the past participle give the **present perfect tense:**

    er hat geschrieben   he has written, he wrote

For some *but not all* intransitive verbs the present perfect is formed with **sein** instead of **haben**:

    er ist geworden   he *has* become, he became
    sie sind gegangen   they *have* gone, they went

Note that the German present perfect tense is equivalent to either the corresponding English tense (he has written) or to the English simple past (he wrote). It is NOT equivalent to HAD written. **Er hat geschrieben** can be translated "He has written" or "He wrote" but NEVER "He had written."

With **werden**:

The present tense forms of **werden** combined with the infinitive give the future tense:

    er wird schreiben   he will (*or* shall) write
    er wird sein   he will (*or* shall) be
    er wird haben   he will (*or* shall) have
    er wird werden   he will (*or* shall) become

The verb **werden** (in any tense) combined with the past participle gives the passive voice*:

es wird verwandelt	it is (*or* gets) transformed
es wird ausgetrieben	it is (*or* gets) driven out
sie werden kontrolliert	they are controlled

(b). *Identifying Past Participles*

In the discussion on Word Order in Lesson 2, Section C, it was shown that past participles serving as part of a verb tense are to be found in last or nearly last position in each clause. In the brief description on verbs in the same lesson, Section B, it was shown that German verbs are either stem-changing (**finden-fand-gefunden**) or non-stem-changing (**zeigen-zeigte-gezeigt**). For the most part, stem-changing verbs have past participles ending in **-en** (**gefunden**), and non-stem-changing verbs have past participles ending in **-t**.

Dictionaries list mainly those past participles involving a stem-change from the infinitive, such as English *spoken* from *speak*.

This means that in the great majority of cases the reader must be able to trace a suspected past participle by himself to the corresponding infinitive before he can look up the latter to get the meaning of the verb.

If there is a **-t** (or **-et**) ending, substitute **-en**†

> kontrolliert → kontrollieren
> verzeichnet → verzeichnen

If there is **also** a **ge-** prefix, remove it:

> gezeigt   → zeigen
> abgemacht → abmachen

If the suspected past participle has a prefix **ge-** and ends in **-n** or **-en**, look it up as it is and the dictionary will give you the corresponding infinitive, which you must then look up:

---

* For details see Lesson 9 B.

† If **l** or **r** precedes, substitute **n**: **verwandelt: verwandeln; gewandert: wandern.**

geschlossen → schließen

If there is a prefix **be-, emp-, ent-, er-, ver-, zer-**, substitute **ge-** and proceed as above. But when you have the infinitive, reattach the prefix and look up the resulting combination*:

entschlossen → geschlossen → schließen → entschließen

Warning: Some verbs, both stem-changing and non-stem-changing, have a permanent **ge-** prefix. For the former the dictionary will show this, for the latter there is nothing to do but try again *with* the ge-prefix if the unprefixed verb does not seem to fit:

gehört may be the past participle of
either **hören** to hear, or **gehören** to belong.

Warning: Past participles are also used as *adjectives* (e.g., *saturated* solution, *broken* glass), in which case they will have *adjective endings* (-e, -em, -en, -er *or* -es) *after* the -*t* or -*en*. Such forms can never be part of a verb tense, but they must still be traced to their infinitive to get their meanings. First remove the adjective ending, then proceed as before:

eine **gesättigte** Lösung: gesättigt → sättigen
zerbrochenes Glas: zerbrochen → gebrochen → brechen → zerbrechen

(c). *Word Order with Compound Tenses*

The auxiliary verb (which is the variable form) comes in *second* position in independent clauses and the constant form comes *last:*

Die Physik **hat** hier große Fortschritte **gemacht**.
Physics has made great progress here.

Das Feuer **ist** allmählich **ausgegangen**.
The fire gradually went out.

---

* There are also a few other inseparable prefixes of less frequency: **miß-, voll-** and **wider-**. Besides, **durch-, über-, um-, unter-** and **wieder-** can sometimes be inseparable prefixes. See Lesson 10, Section C.

Der Wind **wird** einen Druck auf die Fensterscheibe **ausüben**.
The wind will exert a pressure on the window pane.

Ein starker Druck **wird** auf die Fensterscheibe **ausgeübt**.
A strong pressure is exerted on the window pane.

Auxiliaries can also be in *first* position:

Ist das Feuer ausgegangen?
Has the fire gone out?

In *dependent* or *subordinate* clauses the auxiliary is *last* and the constant form immediately precedes it:

Es ist wichtig, daß sie bald **gefunden werden**.
It is important that they *be found* soon.

. . . , wenn ein starker Druck **ausgeübt wird**.
. . . when a strong pressure *is exerted*.

E. NON-TENSE-FORMING AUXILIARIES

A group of auxiliary verbs that do not serve to form tenses of other verbs, but have meanings of their own, are used with the infinitive. All of them have only two instead of the usual three forms for the present tense. They are usually called "Modal Auxiliaries." See Lesson 11 C.

| ich, er, sie, es | kann | } *I, he,* etc., *can* |
| wir, Sie, sie | können | |

| ich, er, sie, es | muß | } *I, you,* etc., *must, have to* |
| wir, Sie, sie | müssen | |

| ich, er, sie, es | soll | } *I am supposed to, I am to,* etc. |
| wir, Sie, sie | sollen | |

Hier **soll** x eine rationelle Zahl darstellen.
Here x *is supposed to* represent a rational number.

Das Problem, das hier genauer untersucht werden **soll** . . . .
The problem that *is to* be more closely investigated here . . . .

Das Feuer **kann** allmählich ausgehen.
The fire *can* gradually go out.

..., da das Feuer bald ausgehen **muß**.
... since the fire *must* soon go out.

Das Feuer **muß** ausgegangen sein.
The fire *must* have gone out.

F. DETACHED VERB-PREFIXES

English verb-phrases like *put on* and *take off* can often be separated by the insertion of their objects:

We *took* our coats *off*.
We *put* our hats *on*.

Similarly in German, the adverbial part of such a combination (*off, on*, etc.) is often separated from the verb. In fact, this adverbial part (called a *separable prefix*) appears at the *end* of the clause if the verb (*take, put*, etc.) is in first or second position. Thus if adverbial words like **auf, an, zu, zurück** appear at the end of a clause, they are to be construed as prefixes belonging to the verb that appears in first or second position in the clause:

Das Feuer **ging** allmählich **aus**.
The fire gradually went out.

Führen wir diese Gleichung auf ihre einfachere Form **zurück**.
Let us trace this equation back to its simpler form.

Solche Variationen **treten** in fast allen Fällen **auf**.
Such variations occur in almost all cases.

To find such verbs in the dictionary, attach the separable prefix before looking them up:

ausgehen	to go out
zurückführen	to trace back
auftreten	to occur

EXERCISE
ELASTIZITÄT UND FESTIGKEIT

Steht ein fester natürlicher Körper unter Einfluß äußerer Kräfte, so bringen diese eine mehr oder weniger meßbare Verschiebung der gegenseitigen Lage der kleinsten Teile des Körpers und eine Veränderung der sichtbaren Gestalt desselben[1] hervor. Hören jene Kräfte auf [2] zu wirken, so nimmt der Körper in verschieden hohem Grade seine ursprüngliche Gestalt wieder an. Die neue Gestalt kann auch eine bleibende[3] geworden sein.

Die Eigenschaft der natürlichen Körper, nach Aufhören der Wirkung dieser Kräfte bis auf einen gewissen Rest die ursprüngliche Gestalt wieder anzunehmen, heißt **Elastizität**.

Körper, bei welchen dieser Rest = 0 wäre,[4] heißen **vollkommen elastisch**, solche, bei welchen von einem Punkte des Körpers ausgehend [5] die Elastizitätsverhältnisse nach allen Richtungen dieselben sind, **isotrop**. Vollkommen elastische Körper gibt[6] es in der Natur nicht.

Der Höchstwert des Widerstandes, den ein Körper der Zerstörung seines Zusammenhanges entgegenzusetzen vermag, heißt seine **Festigkeit** (**Bruchfestigkeit, Tragfestigkeit, Bruchmodul**) (vgl.[7] 2, Anm.[7] 4, Schluß).

Je nach Art der Wirkung der äußeren Kräfte unterscheidet man (unter Voraussetzung[8] stabförmiger Körperform):

1. die **Zugfestigkeit** oder den Widerstand gegen **Zerreißen**.
2. die **Druckfestigkeit** oder den Widerstand gegen **Zerdrücken**.
3. die **Biegungs-**[9] (**relative**) **Festigkeit** oder den Widerstand gegen **Biegung**.
4. die **Schub-(Scher-)Festigkeit** oder den Widerstand gegen **Abscheren**.
5. die **Knickfestigkeit** oder den Widerstand gegen **Zerknicken**.
6. die **Drehungs-(Torsions-)Festigkeit** oder den Widerstand gegen **Verdrehung** (Torsion).
7. die zusammengesetzte **Festigkeit** oder den Widerstand gegen **Beanspruchungen**, die aus mehreren der zuvorgenannten zusammengesetzt sind.

Die Ermittelung der Zahlwerte der Festigkeit ist Gegenstand von Versuchen der Ingenieurlaboratorien (Wedersche[10] Festigkeitsmaschine).

Bei manchen Materialien, z.B. Eisen, steht die Festigkeit wesentlich in Zusammenhang mit der chemischen Zusammen-

Exercise 3

setzung. Innerhalb gewisser Grenzen erhöht die Zunahme des Kohlenstoffgehalts des Eisens dessen[11] Festigkeit gegen ruhende Last, vermindert sie aber gegen Stöße.

Adapted from *Festigkeitslehre* by W. Hauber, pp. 7–10, Sammlung Göschen, Walter de Gruyter & Co., Berlin, 1908.

Notes for Exercise

[1] **Desselben**, of it (literally, of same). See Appendix 1.

[2] **Auf** is a prefix belonging with the verb **hören; zu wirken** is an "infinitive clause." Short infinitive clauses are not set off by commas, but clauses that introduce them are generally closed before the beginning of the infinitive clause: Hören jene Kräfte auf zu wirken

[3] **Eine bleibende**, a lasting one. The noun **Gestalt** is understood after this adjective, hence the ending.

[4] **Wäre**, would be, short conditional of the verb **sein.**

[5] **Von . . . ausgehend**, starting from. . . . The **-end** ending (infinitive + d) is equivalent to the English *-ing*, with the proviso that the German forms can be used only as adjectives or adverbs.

[6] **Es gibt** (+ accusative case) (idiom), there is, there are.

[7] **Vgl.** = **vergleiche**, compare, cf. **Anm.** = **Anmerkung**, note.

[8] **Unter Voraussetzung**, assuming. Phrases with **unter** and a noun ending in **-ung** are often equivalent to the *-ing* form in English.

[9] The hyphen is meant to indicate that **Biegungsfestigkeit** is one word.

[10] **Werdersche.** Family names are turned into adjectives by means of a suffix **-sch**, to which adjective endings can then be attached: "Werder's strength-testing machine."

[11] **Dessen**, its. See Appendix 1.

LESSON **4**

The Noun-Adjective System

A. CASES—ARTICLES AND LIMITING ADJECTIVES

The type of variation found in English in personal pronouns, as for example:

$$\left.\begin{array}{l}\text{he}\\\text{she}\\\text{we}\end{array}\right\} \text{subject-form} \quad \text{vs.} \quad \left.\begin{array}{l}\text{him}\\\text{her}\\\text{us}\end{array}\right\} \text{object-form}$$

is also found in German for the definite article and limiting adjectives, i.e., those modifiers that limit the range of meaning of the noun: "this, that, every, any, some, much, many, my, your, his," etc. While the noun also undergoes some variation in form, it is chiefly the article or limiting adjective modifying it that is subject to change.

Instead of referring to *he, she, we,* etc., as subject-forms, the technical term is to say that they are *in the nominative case,* while those forms serving as objects are said to be in the *accusative case,* etc.

In German there are four cases for each gender and four cases for the plural. The following table shows the variations of the definite article. Technically such a set of variations for gender, number, and case is called a *declension*.

TABLE 6. DECLENSION OF THE DEFINITE ARTICLE

CASES	MASCULINE	FEMININE	NEUTER	PLURAL
Nominative	der	die	das	die
Genitive	des	der	des	der
Dative	dem	der	dem	den
Accusative	den	die	das	die

It will be noted that in actual fact there are only six forms, each of which can be a number of different case-forms or gender-forms:

der masculine nominative (the); feminine genitive (of the), & dative (to the); plural genitive (of the)
die feminine & plural nominative & accusative (the)
das neuter nominative & accusative (the)
dem masculine & neuter dative (to the)
den masculine accusative (the); plural dative (to the)
des masculine & neuter genitive (of the)

From the reader's point of view it is important to recognize these forms and be aware of the possible function of each. Thus **die** and **das** can only modify nouns used as subjects (nominative) or direct objects (accusative), **des** can only be genitive singular and will usually be translated *of* or *of the*, etc.

In order to recognize the forms **das, dem,** and **der** (dative), the reader must realize that these are often contracted with a preceding preposition. Thus **ins** stands for **in das, im** for **in dem, vom** for **von dem, zum** for **zu dem**; **der** enters into only one contraction, namely **zur** (= **zu der**). See Lesson 13.

1. *The Four German Cases and Their Uses*

(*a*). The *nominative* case is the *subject* form as well as that of predicate nouns:

Was ist **der Sinn** dieser Zusammensetzung? (*Subject*)
What is the point of this combination?

Diese Tatsache ist **der Beweis**, den wir brauchen (*predicate noun*).
This fact is the *proof* that we need.

(*b*). The *genitive case* includes possession and many similar relationships, most of which are expressed in English by using the preposition *of*. Note its characteristic endings, especially the **-s** or **-es** added to most masculine and neuter nouns:

das Öhr **der Nadel**	*the needle's eye, the eye of the needle*
der Verfasser **des Artikels**	the author *of the article*
an Stelle **der** alten **Motoren**	in place *of the* old *motors*
trotz **des Wetters**	in spite *of the weather*
des Morgens	in the morning

(*c*). The *dative case* is primarily the case of the indirect object, but also includes other relationships involving the prepositions *to* and *for* in English:

Wir geben (senden, bringen, zeigen) **dem Forscher** die Zeitschrift.
We give (send, bring, show) *the scientist* the periodical (or the periodical *to the scientist*).

Die Anstalt kaufte (besorgte, eröffnete) **dem Forscher** ein neues Laboratorium.
The institution bought *the scientist* a new laboratory.
The institution bought (provided, opened) a new laboratory *for the scientist*.

der anderen Substanz ähnlich
similar *to the other substance*

Es entspricht nicht **den Tatsachen.**
It does not correspond *to the facts.*

(*d*). The *accusative case* is the direct object form but is also used for expressions of definite time, where it often replaces the English preposition "for":

Den anderen Vorgang beobachteten die fremden Gäste später.
The foreign visitors observed *the other process* later.*

Er arbeitete **die ganze Nacht (den ganzen Tag, das ganze Jahr)** daran.
He worked on it *all night (all day, all year)*.

Wir haben **drei Stunden (einen ganzen Tag, viele Jahre)** gewartet.
We waited *for three hours* (waited *one whole day, for many years*).

2. Cases after Prepositions

In addition to the uses briefly outlined above, the genitive, dative, and accusative cases each regularly appear after certain sets of *prepositions:*

durch diesen neuen Apparat by means of this new apparatus	(accusative)
in das kalte Wasser into the cold water	(accusative)
in dem kalten Wasser in the cold water	(dative)
wegen des schlechten Wetters because of the bad weather	(genitive)
aus anderen Gründen for other reasons	(dative)

For details and lists of prepositions "governing" each case, see Lesson 13 and Appendix 2

3. Dieser-*words*

The definite article is a member of a larger category of modifiers called *limiting adjectives*. Within this category are a number

---

* Note that this sentence begins with the object. It is a legitimate device, in order to avoid excessive rearrangement of parts of the sentence, to use a passive voice translation: *The other process was observed by the foreign guests later.*

of modifiers with a declension following that of the definite article. They are customarily listed in their masculine singular nominative form, i.e., corresponding to the masculine **der:**

aller	all, any
dieser	this (*plural* these), the latter
jeder	each, every, any (*no plural*)
jener	that (*plural* those), the former
mancher	many a, (*plural*) **manche** some*
solcher	such
welcher	which

These so-called **dieser**-words have the same declension as the definite article if the -e of **die**, the -er of **der**, etc., are considered endings, and the -as of **das** is replaced by -es:

das	:	alles	dieses	jedes	jenes	manches, etc.
dem	:	allem	diesem	jedem	jenem	manchem, etc.
den	:	allen	diesen	jeden	jenen	manchen, etc.
der	:	aller	dieser	jeder	jener	mancher, etc.
des	:	alles	dieses	jedes, etc. (with genitive case interpretation)		
die	:	alle	diese	jede, etc.		

Thus if **die neue Methode** means "the new method"
then **diese neue Methode** means "this new method"
  jede neue Methode means "every new method"
and **welche neue Methode** means "which new method," etc.

And if **die Genauigkeit der Messung** means "the accuracy *of the* measurement," **die Genauigkeit aller Messung** must mean "the accuracy *of all* measurement," and **die Genauigkeit jener Messung** must mean "the accuracy *of that* measurement," etc.

Additional uses of these **dieser**-words and other "demonstratives" are given in Appendix 1.

Another group of limiting adjectives with a uniform declension includes the indefinite article **ein** (a, an, one), the negative adjective **kein** ("no," as in "no money"), and the possessive adjectives.

---

* In the restrictive sense of the word *some*, i.e., "some, but not others."

Since they sometimes occur with no ending at all, they are listed in their endingless form:

mein	my
sein	his, its (referring to a masculine or neuter possessor)
unser	our (the *-er* is part of the stem of the word)
ihr	her, its (referring to a "feminine" possessor), their
Ihr	your

The endingless form occurs as

masculine nominative singular

ein ander**er** Weg	as compared to	**der** ander**e** Weg
another method		dies**er** ander**e** Weg
		jed**er** ander**e** Weg, etc.

neuter nominative and accusative singular

ein selten**es** Element	as compared to	**das** selten**e** Element
a rare element		dies**es** selten**e** Element
		jed**es** selten**e** Element, etc.

Otherwise these so-called **ein**-words have the same endings as **dieser**-words and can be relied on to have the same grammatical significance:

> ein**es** selten**en** Elements   von kein**em** Einfluß
> of a rare element   of no influence
>
> sein**er** neuen Theorie
> of ⎫
> to ⎭ his new theory
>
> (could be *genitive* or *dative*)

All of the above limiting adjectives, both **dieser-** and **ein-**words, can be used as *pronouns*. In that case, however, even the otherwise endingless forms of the **ein-**words appear with endings (**-er** for masculine, **-es** for neuter):

Von allen Versuchen gelang nur dies**er**. (= dieser Versuch)
Of all the attempts only *this one* was successful.

Von allen Versuchen gelang nur ein**er**. (= ein Versuch)
Of all the attempts only *one* was successful.

Eines der bekanntesten Elemente.
One of the best known elements.

Von den beiden Resultaten ist das eine positiv, das andere negativ.
Of the two results (the) *one* is positive, the other negative.

B. ADJECTIVES AND ADVERBS

German adjectives appear in two forms:

1. With an Ending

This is the *attributive* form and regularly precedes the noun it modifies (unless the latter is left understood). The possible endings are **-e, -em, -en, -er, -es**:

    ein neues Exemplar   a new copy
    wichtige Ergebnisse   important results
    in warmem Wasser   in warm water

Note. Adjectives with stems ending in **-el** drop the **e** before the **l** when endings are added: **edel** noble, **edle Metalle** noble metals. Some adjectives whose stem ends in **-er** drop an **-e** when an ending is added: **andern** or **andren = anderen**. The adjective **hoch** (high) changes its stem to **hoh-** when endings are added: **hohe Preise** high prices.

2. Without an Ending

This is the *predicate* form and usually comes at the end of a clause or just before a last-position verb:

Dieses Exemplar ist **neu.**
This copy is *new.*

Diese Ergebnisse sind **wichtig.**
These results are *important.*

Das Wasser darf in solchen Fällen nicht zu **warm sein.**
The water must not be too *warm* in such cases.

Adjectives and Adverbs

The endingless form of the adjective also serves as the adverb:

Adjective: die **gewöhnliche** Methode
the *usual* method

Adverb: die **gewöhnlich** angewandte Methode
the *usually* applied method

es geht **gewöhnlich** schneller
it *usually* goes faster

seine Erklärung ist **unnötig** kompliziert
his explanation is *unnecessarily* complex

seine **unnötig** komplizierte Erklärung
his *unnecessarily* complex explanation

Dictionaries list adjectives in the endingless form and do not give separate translations for the corresponding adverb unless the latter has special meanings not obviously derivable from the adjective:

vortrefflich (*adj.*) (excellent) The formation of the adverb "excellently" is left to the reader.

ganz (*adj.*) (whole, entire) has adverbial uses not derivable from its adjective uses, so the dictionary will list these, e.g., very; rather, fairly.

For reading purposes, it is less important to know exactly what grammatical ending (-e, -em, -en, -er, -es) an adjective should have in a specific wording than to know whether or not it should have an ending at all. An endingless adjective must never be linked attributively to a noun, even if it stands before it.

Man nennt diesen Vorgang **einfach** Schwingung. (adverb)
We *simply* call this process oscillation.

Dieser Vorgang stellt eine **einfache** Schwingung dar. (adjective)
This process represents a simple oscillation.

Die Größen sind sehr **verschieden**. (predicate adjective)
The magnitudes are very different.

verschiedene große Stücke (attribute adjective)
various large pieces
verschieden große Stücke (adverb)
different-sized pieces (lit.: variously large pieces)

Note that the -en of the adjective **verschieden** is part of the stem; endings are added beyond this -en: verschieden**e**. This situation is particularly frequent in the case of past participles used as adjectives, e.g., **entschieden**, the past participle of the verb entscheiden (to decide):

Das ist **entschieden** richtig.
That is *decidedly* correct.

eine **entschiedene** Verbesserung.
a *decided* improvement.

Any past participle, whether its stem ends in -t or -en, cannot be construed as part of a compound verb tense if it has an ending beyond the -t or -en:

Wir haben außer dem Osazon krystallisiert**e** Hydrazone erhalten.

Besides osazone, we have obtained crystalline hydrazones.

The form **krystallisierte** cannot be combined with **wir haben** to mean "we have crystallized" because (*a*) it has an ending beyond the -t and (*b*) it is not in *last* position.

If an adjective appearing in a position where it should have no ending has an -er attached to the stem, then this -er must be interpreted as the comparative ending (as in English small*er*, fast*er*):

Man kommt auf diese Weise schnell**er** zum Ziel.
We reach our goal faster (*or* more rapidly) this way.

The comparative form may also appear as an attributive adjective, with an ending *after* the -er:

ein rascher Vorgang   a rapid process
ein rascherer Vorgang   a more rapid process

Details on adjective endings and the comparative and superlative of adjectives will be taken up in Lesson 12.

C. RECOGNITION OF NOUN PLURALS

In English most nouns can be made plural by adding -*s* or -*es* (atoms, boxes). In German the **-s** plural is restricted to a very minute number of words of foreign origin usually ending in **-o** or **-a** (**Neutrinos, Kontos, Kameras**) or in French sounds (**Restaurants**), and its occurrence is therefore negligible. For the most part it can be said that *a final* **s** *is not the sign of a plural in German* (it is usually a genitive *singular:* **des** Druckes, eines scharfen Winkels).

The formation of the German plural is a highly complex subject involving many exceptions and special cases. It includes most of the characteristics considered irregular or exceptional in English:

No change	(sheep, series)
Internal change	(tooth-teeth, mouse-mice, man-men)
Change in ending	(radius-radii, phenomenon-phenomena)
Endings other than -s	(ox-oxen, child-children)

For the learner who wishes to speak or write German, noun plurals are a serious problem best solved by the tedious process of learning the plural of each noun individually. But for the reader the problem is less serious and positive identification of a noun as singular or plural is almost always possible if three fundamental principles are kept in mind:

1. It is always easier to recognize a noun as plural if its singular is already known.
2. The noun itself cannot always be relied on alone to show whether it is singular or plural. Often the preceding articles and adjectives provide the significant clue, and if the noun is the subject of a clause, the verb provides the clue.
3. Differences in linguistic usage may require a singular in German where a plural may be used in English or vice versa, but the reader should always know whether the German form is actually a singular or plural:

Kenntnisse (*plural*) knowledge (*singular*)
zwei Stück* Zucker (*singular*) two pieces (*or* lumps) of sugar (*plural*)

Basic Rules

1. Final -s or -es is practically *NEVER* the indication of a plural.
2. A noun of one syllable cannot be a plural. (Druck, Gas, Tag, Zeit, Welt)
3. Umlaut (ä, ö, ü, äu) is often the sign of a plural but is of little real help unless one already knows that the singular form does not have the umlaut:

diese Brücke	this bridge	
diese Stücke	these pieces	(*singular:* Stück)
diese Bürste	this brush	
diese Würste	these sausages	(*singular:* Wurst)

4. The possible endings on plural forms are, in approximate order of frequency:

-en, -e, -er, -eln, -ern, -el

Unfortunately the same endings can also occur on singular forms. Thus additional clues are necessary. There are two large groups of German nouns whose plurals can be regarded as predictable from their singular form. All the rest are arbitrary and unpredictable with respect to the way in which they form their plurals†:

(*a*). Feminine nouns always have a plural different from the singular: **die Faser—die Fasern** (fiber)‡

---

* This is standard practice with masculine and neuter nouns of measure or quantity.

† Many nouns of one syllable, as well as most words of non-German origin stressed on the last syllable, add an **-e** to form their plural. The former often take umlaut, the latter never. Note that forms like **Chloride, Sulfate, Olefine** are plurals in German. The singulars are **Chlorid, Sulfat, Olefin**, etc.

‡ Feminine plurals are by far the easiest to recognize because the vast majority of them are formed by means of an **-n** or **-en** ending and no internal

(b). A large number of masculine and neuter nouns have a plural identical in form to the singular (Specifically, those whose singular ends in -er, -el, -en, -chen, or -lein):

der Körper—die Körper (body)

5. There is no excuse for failing to recognize a subject noun as plural if the variable verb is plural:

Die **Werte bleiben** stets konstant.
The values always remain constant.

. . . da **die Brennpunkte** nicht genau festgestellt werden **konnten.**
. . . since the focal points could not be exactly determined.

6. It is equally obvious that a noun with the specified endings must be plural if it is modified by an obviously plural numeral or quantity expression:

sechs Ecken	mehrere Zweige	einige Stücke
six corners	several branches	a few pieces

7. A noun cannot be plural if it is modified by obviously singular articles or adjectives, especially those ending in -es or -em:

zwei Tage (*plural*)	two days
an dem Tage (*singular*)	on that day
in diesem Paragraphen (*singular*)	in this paragraph
einige Paragraphen (*plural*)	a few paragraphs
jedes Koeffizienten (*singular*)	of each coefficient
alle diese Koeffizienten (*plural*)	all these coefficients

8. The dative plural of all nouns must end in -n. This means that plurals not already ending in -n will appear in the dative case with an -n in addition to any plural ending (except -s):

den Büchern, dative of **die Bücher,** plural of **das Buch** (book)
den Punkten, dative of **die Punkte,** plural of **der Punkt** (point)

---

change (no umlaut). Since enormous numbers of feminine nouns are recognizable by characteristic suffixes such as **-ei, -ie, -heit, -keit, -ität, -ung, -ur, -schaft,** etc., there is no excuse for failing to recognize their plurals (formed by adding **-n** or **-en**): **Einheiten, Ablesungen, Eigenschaften, Temperaturen.**

Besides, all *modifiers* that can take endings will also end in **-n** or **-en** in the dative plural:

in all**en** dies**en** hier angeführt**en** linear**en** Funktion**en**
in all these linear functions cited here

It will be seen from the foregoing that the German noun itself does not actually undergo any great *variation for case*, while it may or may not be subject to change in forming the plural. As an aid to identifying nouns and their few variations, many dictionaries indicate the "principal parts" of each noun, namely the genitive singular and the nominative plural, in the following form:

Körper, *m* (-s, —)

The -s indicates that the genitive singular is **des Körpers**.
The — indicates that there is no change for the plural.

Baum, *m* (-es, ⁔e)

The ⁔e indicates that the plural is **Bäume**.

Nouns whose genitive is formed with the **-s** or **-es** usually involve no other changes in the singular, i.e., the dative and accusative singular are identical to the nominative (**Körper, Baum**), but the dative singular of nouns with the stress on the last syllable (including monosyllables) may add an **-e**: **dem Gehalte, dem Punkte, dem Baume**. Masculine nouns whose genitive does NOT end in **-s** or **-es** belong to a special group for which *der Student* may serve as a model:

STUDENT m (-en, -en)

	SINGULAR	PLURAL	
Nominative	der Student	die	⎫
Genitive	des Studenten	der	⎬ Studenten
Dative	dem Studenten	den	⎪
Accusative	den Studenten	die	⎭

D. COMPOUND NOUNS

In English we readily combine two nouns into one in such combinations as *sunshine* (sun & shine), *bedroom, toenail* and *dishwater*. We also speak of *ice water, egg rolls, reception rooms* and *coat hangers,* and even though we don't write these combinations as one word, we recognize them as units no less than *dishwater, bedrooms* and *sunshine*. Nor do we need to stop at two component nouns. We can say *evening sunshine, bedroom furniture, coat hanger wire* and *wire coat hanger,* and we can go even further:

bedroom furniture dealer
bedroom furniture dealers' association
bedroom furniture dealers' association secretary

German nouns can be strung together in exactly the same way, with this important difference in spelling: they are always combined into one word:

Lebensversicherungsgesellschaft
life insurance company

Schlafzimmermöbelhändlerverein
bedroom furniture dealers' association

Note. There has been a growing tendency toward breaking up combinations of more than two nouns by using a hyphen: **Harnstoff-Additionsverbindung** instead of **Harnstoffadditionsverbindung** (urea addition compound). Still, many authors continue to adhere to the older practice.

The main difficulty in reading such compounds is to identify the individual nouns that make them up. Except for those of very great frequency (e.g., **Lebensversicherung**), dictionaries do not assume the burden (and the bulk) of listing such strings of nouns. The reader must identify the individual components and analyze the combined meaning for himself.

The word **Leben s versicherung** will illustrate the frequency of the genitive-case -s ending in such combinations. This **s** has come to be interpreted as a sort of link rather than an ending and is therefore often added to feminine components as well:

Leben s versicherung s gesellschaft

(**Versicherung**, ending in **-ung**, is obviously a feminine noun). Less frequently, feminine components will be encountered with an **-n** or **-en** ending:

Raketenschiff rocket ship   (**die Rakete**)

Note that, whether or not any such endings are present, it is not always possible to string the components together in the same order in English:

Lichtgeschwindigkeitsbestimmung
(*Literally*) light speed determination
i.e., determination of the speed of light

The gender of any compound noun is that of the *last* component:

**das Leben + die Versicherung
die Lebensversicherung**

**die Regierung + das Mitglied
das Regierungsmitglied** (member of the government)

When two compounds have the last component in common, repetition is avoided by mentioning the last component only for the second compound. In writing, a hyphen then follows the detached first component:

Brandschaden- und Lebensversicherung
(fire   and life insurance)

instead of:

Brandschadenversicherung und Lebensversicherung

Warning. If two German nouns stand side by side in a sentence, there is usually no grammatical relationship between them,* and they cannot be considered a unit. Two nouns forming a unit (the first acting as a modifier of the second) are always spelled as one word in German:

. . . wenn unter Einfluß der beweglichen **Last Zug** mit Druck abwechselt.

When, under the influence of the movable *load, tension* alternates with compression.

Last is part of the phrase   **der beweglichen Last**
of the movable load

---

* The chief exception is the wording in quantity expressions such as **zwei Glas Wasser**, two glasses of water.

Zug is part of the subject of the clause   **wenn . . . Zug mit Druck abwechselt**   when tension alternates with compression

Under no circumstances must the reader jump to the rash conclusion that **Last Zug** could be construed as **Lastzug,** freight train!

EXERCISE
DIE KRISTALLE

(Try carefully to identify all nouns in this article as singular or plural.)

Ein Kristall ist ein homogener Körper, der bei freier Entwicklungsmöglichkeit von ebenen Flächen bestimmter geometrischer Form gesetzmäßig begrenzt ist. Diese gesetzmäßige Flächenanordnung wird durch die gesetzmäßige innere Anordnung der Atome, die den Kristall aufbauen, bestimmt. Die äußere Eigenform eines Kristalls kann zwar im Falle eines Kristallaggregates unentwickelt sein, der gesetzmäßige innere atomare Aufbau wird hierdurch jedoch nicht beeinflußt.

Kristalle und Kristallaggregate entstehen beim Übergang eines Stoffes aus dem gasförmigen oder flüssigen Zustand in den festen Zustand.

In einer nichtkristallisierten oder amorphen Substanz, z.B. Glas treten die Atome in Molekülgruppen ohne gesetzmäßige symmetrische Anordnung auf. Die *Interferenzerscheinungen der Röntgenstrahlen* haben gezeigt, daß die Moleküle beim Übergang einer amorphen Substanz in eine kristallisierte im allgemeinen ihre Bedeutung verlieren.

Begrenzungsstücke der Kristalle. Kristalle werden von Flächen, Kanten und Ecken begrenzt. Die *Kristallflächen* sind meist ebene *Flächen.* Unter gewissen Bedingungen können jedoch auch gebogene Flächen auftreten, z.B. bei den rhomboedrischen Karbonaten, wie Dolomit und Eisenspat.

Zwei benachbarte Flächen eines Kristalls schneiden sich[1] in einer Kante. Mehrere Kanten treffen sich[1] in einer Ecke. Ein

---
[1] These are reciprocal verbs. The pronoun **sich** literally stands for "each other," but this can be omitted in English: Two . . . faces . . . *intersect.* . . . Several edges *meet* in a corner.

Würfel hat z.B. sechs Flächen, zwölf Kanten und acht Ecken. Flächen, die sich in parallelen Kanten schneiden, liegen in einer *Zone*. Der Winkel zwischen zwei Flächen eines Kristalls wird Flächenwinkel, der Winkel zwischen zwei Kanten Kantenwinkel genannt. Die Werte dieser Winkel bleiben für gleichwertige Flächen und Kanten derselben Mineralart stets konstant. Kristallwinkel werden mit dem Goniometer gemessen.

From *German-English Geological Terminology* by W. R. Jones and A. Cissarz, pp. 62-72, George Allen & Unwin Ltd., London, 1931.

LESSON 5

Pronouns

A. PERSONAL PRONOUNS

1. Subject and Object Pronouns

The German subject and object pronouns for persons and things vary as radically as do the corresponding English forms (he-him, we-us, etc.). Here are the most frequently encountered forms:

er	(he, it)	corresponds to	**der**-nouns
es	(it)	corresponds to	**das**-nouns
ihm	(to him, to it)	corresponds to	**dem**, hence is dative singular
ihn	(him, it)	corresponds to	**den** as an accusative singular
sie	(she, her, it, *plural* they, them)	corresponds to	the article **die** and is therefore singular or plural, nominative or accusative.
ihr	(to her, to it)	corresponds to	**der** as a dative singular.
ihnen	(to them) is always dative plural.		

Warning

(*a*). Do not confuse the object pronoun **ihr** with the possessive adjective **ihr**. The latter precedes nouns as an attributive adjective and is variable in that it can take endings:

Object Pronoun: Das verleiht **ihr** einen neuen Charakter. That imparts a new character *to it* (*to her*).

Possessive Adjective: ihr veränderlicher Charakter. (*genitive:* **ihres** veränderlichen Charakters, etc.) its (her, their) variable character

(*b*). Do not confuse **ihnen** with **ihren**. The latter is a variation of the possessive adjective **ihr**.

See also what was said about subject pronouns in Lesson 3 C. Regarding the use of **es** in impersonal expressions, see Lesson 9 C.

2. Reflexive Pronouns

The form **sich** is a *reflexive pronoun* literally meaning "oneself, himself, herself, itself, yourself, yourselves, or themselves."

It can be dative or accusative. In the latter case it serves as the direct object of the reflexive verb. Reflexive verbs can rarely be translated literally (**sich beherrschen** to control oneself); usually the combination has some idiomatic translation:

    zeigen   to show
   Es zeigt sich   it appears, it becomes evident

The pronouns **mir** (dative) and **mich** (accusative) and the pronoun **uns** (dative or accusative) can also occur as reflexive objects meaning "myself" and "ourselves" respectively:

Ich dachte es mir anders.
I imagined it differently.
(*Literally:* I thought of it for myself . . .).

For details on reflexive verbs, see Lesson 9 A.

3. The Subject Pronoun man

In order to make a generalized statement, often equivalent to using a passive, German uses a special subject pronoun **man**. It closely approximates French **on**. There is no universal English translation other than the word *one*, which often sounds artificial if not stilted. Other possibilities are *we, they, people, a person*, or translation by means of a passive. Note that the German word **man** is always a singular, is accompanied by the **er-sie-es** form of the verb, and is *always a subject:*

Man sagt, es wird regnen.

They say ⎫
People say ⎬ it will rain.
It is said ⎭

Man hat auf diesem Gebiet in den letzten Jahren große Fortschritte gemacht.

We have made great progress ⎫
They have made great progress ⎬ in this field in the past years.
Great progress has been made ⎭

If **man** occurs with a verb ending in **-e**, and this **-e** is not part of the past-tense or short conditional ending **-te**, then the sentence is to be interpreted as a formal imperative:

Man zähle bis zehn.   Count to ten.

B. EXAMPLES OF PERSONAL AND REFLEXIVE PRONOUNS

Dieser Wert ist veränderlich. **Er** kann steigen oder sinken.
This value is variable. *It* can rise or fall.

Wenn diese Metalle mit der Säure in Berührung kommen, zersetzen **sie sie**.
When these metals come in contact with the acid, *they* decompose *it*.

Der Vektor und der **ihn** abbildende Pfeil
the vector and the arrow representing *it*

Die Gleichung und die **ihr** entsprechende Kurve
the equation and the curve corresponding *to it*

Hier sind zwei Punkte. **Wir** bezeichnen den einen von **ihnen** mit *a*, den anderen mit *a'*.
Here are two points. *We* designate one of *them* as *a*, the other as *a'*.

Man kann **sich** leicht davon überzeugen.
One can easily be convinced (*literally* convince *oneself*) of it.

Wir können **uns** leicht davon überzeugen.
We can easily be convinced (*literally* convince *ourselves*) of it.

Dies ermöglichte **es ihm**, weiterzuforschen.
This made *it* possible *for him* to continue investigating.

Wenn alle Bedingungen so festgelegt werden, wie wir **es** oben getan haben . . .
If all conditions are established in such a way as we have done (*it*) above . . .

Es wird **sich** zeigen, daß diese Dinge als Zahlen angesehen werden können.
It will become evident (*literally,* show *itself*), that these items can be regarded as numbers.

Special Note. The word **es** introducing a sentence is not always a pronoun in the strict sense. It is often a so-called expletive similar to English *there* in sentences like:

"There came a time . . ." German: **Es kam eine Zeit** . . .

For details, see Lesson 9 C.

C. PREPOSITIONAL COMPOUNDS (**Da**-forms)

Darin liegt der Unterschied.	*Therein* (or *in that*) lies the difference.
das Haus und die Menschen **darin**.	The house and the people *in it* (or *therein*).
die Häuser und die Menschen **darin**.	The houses and the people *in them* (or *therein*).

die Strafe **dafür** — The penalty *for it* (or in legal language: *therefor*).

1. Such **da**-compounds exist for almost all prepositions governing the dative and accusative cases (see Appendix 2). While their literal English equivalents (*therein, therefor, thereof,* etc.) usually have a very formal or legal tone, the German words are a part of everyday language and are best translated by the less formal phrases *in it, in them, in that,* etc. (Note that the **da** of the German word corresponds to the *it* or *them* or *that* of the English phrases.) A few of these **da**-compounds have taken on special meanings or require idiomatic translations:

dabei	in this connection, in this case, in doing this (It does NOT mean *thereby*)
damit	(introducing a dependent clause with the verb(s) in last position) so that, in order that
daher, darum	therefore, for that reason, that is why
dagegen	on the other hand, on the contrary
daneben	beside(s)
darauf	*often means* thereupon
dadurch	thereby

2. Practically all of the **da**-forms can also be used as "separable prefixes" attached to verbs. See Lesson 10 B.

3. *The Anticipating* da-*Form*

The **da**-forms are often used to introduce two kinds of dependent clauses:

(A). Daß-Clauses

Die Verwandtschaft liegt **darin, daß** beide Aufgaben von derselben Gleichung abhängen.

The relationship lies *in (the fact)* that both tasks depend on the same equation.

Die Methode zeichnet sich besonders **dadurch** aus, **daß** keine lästigen Nebenprodukte anfallen.

The method is especially distinguished *by the fact that* no bothersome by-products accumulate.

The above examples suggest that it is expedient to use the words "the fact that" preceded by the preposition in question (*in* or *by*, etc.) for the combination "da-, daß." This will not always give the best English translation but may serve as a starting point for a preliminary rendition which can be smoothed over later:

Das Verfahren von Solvay besteht **darin, daß** Steinsalzlösung mit einer Lösung von Ammonium-Bikarbonat umgesetzt wird zu . . .

Preliminary translation:

The process of Solvay consists *in the fact that* rock salt solution is converted with a solution of ammonium bicarbonate into . . .

Smoother version:

The Solvay process consists *in converting* a rock salt solution, etc.

(B). Zu-Infinitive Clauses

Also besteht unsere Aufgabe **darin,** diese Werte als Funktion von *x* **darzustellen.**

Thus our task lies *in representing* these values as a function of *x*.

Es handelt sich **darum,** eine gültigere Lösung **zu finden.**

It is a matter *of finding* a more valid solution.

Here the best translation is evidently a preposition + the *-ing* form.

4. Hier-*Forms*

Just as, in English, forms like *herein, hereto, hereof, herewith* closely parallel the forms *therein, thereto, thereof, therewith*, so also in German there are forms made up with **hier-**[*] paralleling the **da**-forms, often with little difference in meaning:

---

[*] Sometimes also *hie* before a consonant: **hierbei** or **hiebei, hiervon** or **hievon,** etc.

Hierin liegt die Lösung.
Herein lies the solution.

D. INTERROGATIVES

All German question words begin with **w:**

wann	when	
warum	why	
was	what	**was für (ein)** what kind of (a)
wer	who	
weshalb	why, for what reason	
wie	how	
wieso	how does it come about that	
wieviel	how much, how many	
wo	where	
woher	where from	
wohin	where (to)	
welch—	which, what	

The interrogative pronoun **wer** has case variations:

Nominative	**wer**	who
Genitive	**wessen**	whose
Dative	**wem**	(to) whom
Accusative	**wen**	whom

Corresponding to the English forms such as "whereby, wherein, wherewith," etc., there are **wo**-forms in German:

Worin liegt seine Überlegenheit?
Wherein (*or* in what) does his superiority lie?

Wodurch wird diese Hitze erzeugt?
Whereby (*or* by what means) is this heat produced?

Wovon lesen wir?
What are we reading about?

All of the question words, including the **wo**-forms, can be used to introduce dependent clauses with the verbs in last position. The **wo**-forms often replace a prepositional phrase whose object is a relative pronoun:

Es ist unsicher, wann das geschah.
It is uncertain when that happened.

Er weiß, wovon er spricht.
He knows whereof he speaks (*or* what he is talking about).

Die Annahme, worauf die Theorie beruht. . . .
The assumption on which the theory is based. . . .

EXERCISE A

Unterscheiden sich zwei komplexe Zahlen nur durch das Vorzeichen ihrer zweiten (kartesischen oder Polar-) Koordinate, so nennt man sie zueinander konjugiert k o m p l e x[1] oder kurz konjugiert. Die entsprechenden Punkte liegen spiegelbildlich zur reellen Achse (s. Fig. 1). Heißt die eine von ihnen a, so bezeichnet man die andere gern mit a'. Unterscheiden sie sich durch das Vorzeichen b e i d e r[1] kartesischen Koordinaten, so nennt man sie zueinander e n t g e g e n g e s e t z t.[1] Heißt die eine a, so bezeichnet man die andere mit $-a$. Die entsprechenden Punkte liegen spiegelbildlich in bezug auf den Nullpunkt (s. Fig. 2), die Vektoren sind gleich lang und parallel, aber von entgegengesetzter Richtung.

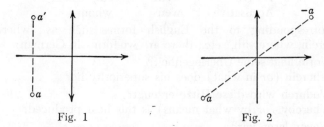

Fig. 1   Fig. 2

Die Aufgabe der folgenden Paragraphen soll nun darin bestehen nachzuweisen,[2] daß die Dinge, von denen wir hier gesprochen haben, im Sinne des §2 Z a h l e n[1] sind. Einen solchen Nachweis scheinen jedoch die letzten Ausführungen dieses §2 grundsätzlich unmöglich zu machen. Dort wurde nämlich[3] gesagt, daß das System der reellen Zahlen (wesentlich) das e i n z i g e¹ System von Dingen sei,⁴ mit denen sich so rechnen läßt,⁵ daß dabei die sämtlichen in §2 zusammengestellten⁶ Grundgesetze der Arithmetik gültig sind. Das ist allerdings der Fall. Wir werden aber sehen, daß nach einer einzigen geringen Modifikation an den Grundgesetzen jener Nachweis dennoch erbracht werden kann. Diese wird darin bestehen, daß man bei den Grundgesetzen der Anordnung nicht mehr verlangt, daß zwischen je zweien⁷ unserer komplexen Zahlen immer eine der drei Beziehungen $<$, $=$ oder $>$ besteht, sondern nur fordert, daß zwischen ihnen eine der beiden Beziehungen $=$ oder $\neq$ besteht. Die O r d n u n g ¹ der komplexen Zahlen ist also eine grundsätzlich andere als die⁸ der reellen Zahlen.

Elemente der Funktionentheorie von Prof. Dr. Konrad Knopp, Tübingen, Sammlung Göschen, Band 1109, Berlin, 1949, pp. 24–25, Walter de Gruyter & Co.

Notes for Exercise A

[1] Letter-spacing is the German equivalent of italics, and is used for emphasis.

[2] **nachzuweisen** is an infinitive clause. The clause introducing it ends with the word **bestehen.**

[3] **Dort wurde nämlich gesagt . . .** = **Denn dort wurde gesagt . . .** For there it was said . . .

[4] Quoting-form of **sein**; translate "was."

[5] **mit denen sich so rechnen läßt. . . .** with which one can figure in such a way. . . . (**sich lassen** often is equivalent to "can be"; here the literal translation would be: "which let themselves be figured with")

[6] **die . . . Grundgesetze** (extended modifier construction). Translate by putting the adjective **zusammengestellt** and its adverbial phrase **in §2** after the noun in English.

[7] between each two. The numbers **zwei** and **drei** are occasionally used with an -en ending in the dative case.

[8] **die** stands for **die Ordnung**: "that of real numbers"

EXERCISE B

The following interrogative sentences are based on statements in Exercise **A**. Translate them and identify the statements in Exercise **A** that would answer these questions in German.

1. Wie nennt man zwei komplexe Zahlen, wenn sie sich nur durch das Vorzeichen ihrer kartesischen Koordinate unterscheiden?
2. Wo liegen die entsprechenden Punkte?
3. Womit bezeichnet man die andere Koordinate, wenn die eine a heißt?
4. Worin soll nun die Aufgabe der folgenden Paragraphen bestehen?
5. Welchen Nachweis scheinen die letzten Ausführungen des §2 unmöglich zu machen?
6. Was wurde dort gesagt?
7. Was werden wir aber sehen?
8. Worin wird diese Modifikation bestehen?

LESSON 6

Word Order

It was stated in Lesson 2 that German word order is largely a matter of finding the verb, or more specifically, the *variable* verb. In fact, the three main classifications of German word order are functions of the three possible positions of the variable verb.

A. VERB FIRST

1. Yes-or-No Questions
   Kann man das Atom zertrümmern?
   Can one smash the atom?
   Nehmen wir diese Dinge an?
   Do we assume these items?

2. Commands (rare in scientific writing)
   Nehmen Sie folgendes an!
   Assume the following.

3. "Let-us" Imperatives
   Nehmen wir einstweilen folgendes an.
   Let us assume the following for the time being.

4. If-less if-Clauses

Sollten diese Änderungen vorgenommen werden,
so würden viele Unannehmlichkeiten beseitigt werden.

If these changes were undertaken (*or* Should these changes be undertaken), many inconveniences would be eliminated.

The **so** of the concluding clause, sometimes replaced by **dann**, serves as a further means of identifying the if-less if-sentence and to distinguish it from other verb-first patterns:

Unterscheiden sich zwei komplexe Zahlen nur durch das Vorzeichen ihrer zweiten Koordinate, **dann** nennt man sie konjugiert.

If two complex numbers differ only in the sign of their second coordinate, then we call them "conjugated."

B. VERB SECOND

This is the usual order for ordinary statements.

1. It begins with either the subject or some sentence element other than the subject. In either case the *variable verb comes second;* in the latter case the subject follows the verb.

Die Arbeit	fängt	heute ungewöhnlich früh	an.
(Subject)	(variable verb)	(adverbial expressions)	(verb prefix)

Work is starting unusually early today.

This order, beginning with the subject, is traditionally known as "normal order."

Heute	fängt	die Arbeit	ungewöhnlich früh	an.
(Adverb)	(variable verb)	(subject)	(another adverb, premodified)	(verb prefix)

Today work is starting unusually early.

This order, beginning with something other than the subject, is traditionally known as "inverted order" or "inversion."

2. It is not unusual for a German sentence to begin with its *object,* or with an *indirect object* or other *non-accusative complement:*

Diesen Schluß zogen die Forscher aus den Fehlschlägen früherer Unternehmungen.

The scientists drew this conclusion (*or* This conclusion was drawn by the scientists) from the failures of earlier undertakings.

Jedem Punkt entspricht ein Koordinatenpaar.

To each point corresponds a pair of coordinates.

Eines ähnlichen Verfahrens bedienten sich auch zwei deutsche Forscher.

A similar process was also made use of by two German scientists. (*Literally:* Of a similar process two German scientists also made use.)

3. For the sake of special emphasis, even elements which normally should come last in a clause can sometimes be found first, especially predicate adjectives and past participles:

Wichtig ist hier die Tatsache, daß die Rohstoffe immer teuerer wurden.

Literally: Important here is the fact that . . .
Translations giving similar emphasis in English:

The important fact here is (*or* The important thing here is the fact) that raw materials were becoming more and more expensive.

Umgekommen bei dem Unfall sind Herr und Frau Schmidt sowie zwei noch unbekannte Insassen des zweiten Autos.

Killed in the accident were Mr. and Mrs. Schmidt as well as two still unidentified occupants of the second car.

4. Normally, in a sentence beginning with an element other than the subject, the subject should immediately follow the second-position variable verb. By way of exception, however, object pronouns are always inserted between verb and subject, and adverbs and prepositional phrases can often be encountered there as well.

Obgleich der Prozeß sehr originell war und großen Erfolg hatte, konnte **ihn** der Erfinder nicht patentieren.

Although the process was very original and had a great success, the inventor could not patent *it*.

Zum Kohlendioxyd als Primärprodukt der Kohlenverbrennung führten **bereits 1939** Untersuchungen eines anderen Autors.

Investigations of another author led already in 1939 to carbon dioxide as the primary product of coal combustion.

Unter den in Frage kommenden Metallen macht **die einzige Ausnahme in diesem Sinn** das Magnesium.

Among the metals concerned the only exception in this sense is made by magnesium.

5. Questions beginning with question words or question phrases likewise illustrate the principle of putting the variable verb *second:*

Worin **liegt** seine Überlegenheit?
Wherein does his superiority lie?

In welchem Zusammenhang **wurde** das erwähnt?
In what connection was that mentioned?

(See also Exercise B of Lesson 5.)

6. The verb is still technically in second position if first position is occupied by a subordinate clause. From the reader's point of view, this amounts to finding the variable verb *first* in the independent clause:

Damit die Leute früher nach Hause kommen können, **fängt** die Arbeit jetzt etwas früher **an**.

So that the employees can get home earlier, work now starts somewhat earlier.

7. Non-subordinating conjunctions:

Conjunctions, or clause-connecting words, followed by subject + verb or immediately by the verb, ordinarily present no great problem to the reader:

Man beachte diese Anweisungen genau, **denn** das Verfahren wird hier gefährlich (or: **denn** hier wird das Verfahren gefährlich).

Observe these instructions closely, *for* the procedure becomes dangerous here.

Das Kohlendioxyd wird oft Kohlensäure genannt, **aber** es wäre besser, diesen Namen zu vermeiden.
Carbon dioxide is often called "carbonic acid" (in German), *but* it would be better to avoid this name.

Difficulties arise for the reader when the conjunction is placed somewhere inside the second clause instead of at its beginning:

Das Kohlendioxyd wird oft Kohlensäure genannt, es wäre **aber** besser, diesen Namen zu vermeiden.
Carbon dioxide is often called carbonic acid; it would, *however*, be better to avoid this name.

Kohlendioxyd ist schwerer als die Luft, **also** ist es möglich (or: ‚es ist **also** möglich), es wie Wasser aus einem Gefäß ins andere zu gießen.
Carbon dioxide is heavier than air, *so* it is possible to pour it like water from one vessel to another.
Carbon dioxide is heavier than air; it is, *therefore*, possible, etc.

Notice the semicolons in the above translations, where German has a comma. In German such "comma-splices" between clauses occur even where there seems to be no close connection between the clauses, as below:

Die Leser dieser Zeitschrift werden sich noch erinnern, von dem Buche „XYZ" gehört zu haben, mancher hat sich dieses Werk vielleicht auch schon beschafft.
The readers of this periodical will remember having heard of the book "XYZ"; many a reader may even have bought this work.

Wir wollen vorläufig ganz im Rahmen der Newtonschen Mechanik bleiben, nur die Ausdrucksweise wollen wir der Relativitätstheorie anpassen.
Let us for the time being remain entirely within the framework of Newtonian mechanics; only the manner of expression do we want to adapt to the relativity theory.

C. VERB LAST

1. The following *subordinating conjunctions* introduce dependent (*or* subordinate) clauses, in which, in German, the variable verb will be found *last*.

als*	when,* as
als ob†	as if
bevor	before
bis‡	until
da	since, inasmuch as
damit	so that, in order that
daß	that, the fact that
ehe	before
falls	if, in case
indem§	by virtue of the fact that, by (+ -*ing* verb)
nachdem§	after
ob	whether
obgleich, obwohl	although
seit, seitdem	since (the time), ever since
sobald	as soon as
so daß	so that
sofern	as far as, insofar as
solange	as long as
sooft	as often as
sosehr	as much as
soviel	as much as, as far as

---

* **Als** occurring after a comparative means "than"; as a subordinating conjunction it usually means "when," especially if the verb in the clause is in the past tense.

† **Als ob** is often shortened to **als,** in which case the variable verb follows immediately: **als ob nichts geschehen wäre** or **als wäre nichts geschehen,** as if nothing had happened.

‡ **bis** can also be a preposition (until, up to). See Lesson 13.

§ Do not confuse with **in dem** or **nach dem** (i.e., with the prepositions **in** or **nach** followed by a dative case article or relative pronoun). See Lesson 13.

Verb Last

soweit	as far as
sowie	as soon as
trotzdem	in spite of the fact that
während ‖	while, whereas
weil	because
wenn	if; when, whenever
wie	as, when, how
wie wenn	as if
wo	where; when
wofern	provided
zumal	especially since

2. As stated in Lesson 5 D, all question words and **wo**-words can introduce dependent clauses, that is, can serve as subordinating conjunctions.

Examples

Da die Wicklungen dabei leicht beschädigt werden **könnten**, ist dieser Versuch natürlich nicht zu empfehlen.

Since the windings could easily be damaged in doing this, this experiment is, of course, not to be recommended.

Man kann den Leiter eines Netzwerkes als Strecke darstellen, **indem** man von allen Schaltelementen und Stromquellen **absieht**.

We can represent the conductor of a network as a straight line by disregarding (*literally* by virtue of the fact that we disregard) all switching devices and current sources.

As already pointed out in Lesson 2, if the verb is in a compound tense or consists of an auxiliary plus constant forms, then the variable verb is last and the constant forms immediately precede it. (For one exception see Lesson 11 C, 1.) The verb forms are then assembled at the end of the dependent clause in a sequence that is exactly the reverse of that required in English. (The word order of dependent clauses is often referred to as "dependent" or "transposed" order.)

Es liegt auf der Hand zu fragen, **ob** auch eine Division einer skalaren Größe oder eines Vektors durch einen Vektor als eine

---

‖ **Während** can also be a preposition governing the genitive case and meaning "during."

Operation **definiert werden kann,** die auf ein eindeutiges Ergebnis führt.

It is the obvious thing to ask whether a division of a scalar magnitude or of a vector by a vector *can be defined* as an operation leading to an unequivocal result.

Es konnte gezeigt werden, **daß** durch diesen Zusatz die Geschwindigkeit der Umsetzung wesentlich **erhöht werden müßte.**

It was possible to show (*literally,* it could be shown) that by this addition the rate of the conversion *would have to be* considerably *raised.*

3. Detachable verb prefixes appear attached to their verb forms in subordinate clauses:

A nimmt zu, während B **abnimmt.**

A increases while B decreases.

Das sind die Änderungen des Vektors, wenn wir von einem Punkt zum anderen **übergehen.**

These are the changes in the vector when we go (over) from one point to the other.

4. Many of the regular subordinating conjunctions have special meanings in combination with **auch** and **immer:**

Wo (wie, wann, wieviel, wer) es **auch** sein mag . . .

Wherever (however, whenever, however much, whoever) it may be . . .

(**Immer** may be found in place of **auch** in the above clauses.)

Wenn es **auch** möglich ist . . . **Auch wenn** es möglich ist . . .

Even if (*or* even though) it is possible . . .

So undenkbar es **auch** sein mag . . .

However inconceivable it may be . . . , *or* Inconceivable as it may be . . .

There are also some special combinations with **daß:**

(*a*). **Als daß**

Der Versuch war zu schwierig, **als daß** man ihn an einem Tag hätte durchführen können.

The experiment was too difficult for it to be carried out in one day.

(*b*). **Ohne daß**

Es ist heutzutage praktisch unmöglich eine Wasserstoffbombe probeweise zu sprengen, **ohne daß** die ganze Welt davon weiß.

It is practically impossible nowadays to test-explode a hydrogen bomb, *without* the whole world knowing about it.

5. *"Relative" Clauses* are taken up in Lesson 7.

EXERCISE
DIE KOMPLEXEN ZAHLEN

Man kam oft durch „formal" richtiges Rechnen auf Ausdrücke, in denen Quadratwurzeln aus negativen Zahlen auftraten und die doch „formal" die Bedingungen des betreffenden Problems erfüllten. Solche Ausdrücke bezeichnete man dann als imaginäre, d.h. eingebildete oder unwirkliche Zahlen.

So kam es, daß man die Wurzeln aus negativen Zahlen nicht einfach verwarf, sondern sich ihrer in immer steigendem Maße und mit immer größerem Erfolge bediente, obwohl man ihnen keine unmittelbare Bedeutung zu geben vermochte und ihr Gebrauch daher rätselhaft und unbefriedigend blieb. Der größte Teil der Dinge, die in diesem Bändchen besprochen werden, wurde schon gegen Ende des 17. und im Laufe des 18. Jahrhunderts, insbesondere von L. Euler (1708–1783) gefunden. Aber erst um die Wende des 18. Jahrhunderts begann man hier ganz klar zu sehen. Eine Abhandlung des Landmessers *Caspar Wessel* aus dem Jahre 1797 und ebenso eine solche[1] von J.-R. Argand aus dem Jahre 1806, in denen eine Lösung des Rätsels gegeben wurde, fanden zunächst keine Beachtung. Nicht anders ging es ähnlichen Versuchen[2] einiger weiterer Mathematiker. Erst als C. F. Gauss 1831, unabhängig von seinen Vorgängern, dieselben[3] Auffassungen entwickelte, war die Zeit für das volle Verständnis dieser Dinge reif geworden. In kurzer Zeit, insbesondere durch die rein arithmetisch gehaltene Darstellung von W. R. Hamilton aus dem Jahre 1837—die Arbeiten der vorher genannten Mathematiker stellten die Dinge in geometrischem Gewande dar—, war alles Geheimnisvolle[4] und Rätselhafte[4] an jenen „sinnlosen Ausdrücken" verschwunden, die heute auf Grund einer geklärten Einstellung zu den Grundlagen unserer Wissenschaft keinerlei begriffliche oder tatsächliche Schwierigkeiten mehr machen.

Elemente der Funktionentheorie, pp. 19–21 (*cf.* Exercise 5A).

Notes for Exercise

[1] **eine solche von J.-R. Argand,** one by J.-R. Argand. See Appendix 1.

[2] This is based on the idioms **Wie geht es Ihnen?—Es geht mir gut,** *How are you?—I am well,* but the literal meaning is: *How are you faring?—I am faring well.* "Similar attempts . . . fared no differently."

[3] The same. See Appendix 1.

[4] **alles Geheimnisvolle und Rätselhafte,** everything mysterious and enigmatic. This is the standard form for adjectives after **alles.** *Note also:* **etwas (nichts, viel) Geheimnisvolles,** something (nothing, much that is) mysterious.

LESSON 7

Word Order (Continued)

A. RELATIVE PRONOUNS

1. One cannot read very far in German without encountering the words **der, die, das** and some of their other variations introducing subordinate clauses with the verb(s) in last position. The word **welcher,** in all its variations, is used in a similar manner:

Schließlich haben sich alle Autoren, **die** sich nach uns mit der Lösung dieses Problems befaßt haben, unserer Formulierung angeschlossen.

Finally all authors *who* concerned themselves after us with the solution of this problem, joined us in our formulation.

The relative pronoun **die** is plural to agree with its antecedent **Autoren;** it is the subject of the clause it begins, hence is in the nominative case.

Das Gefäß, in **das** das Wasser fließt, muß rein sein.

The vessel into which the water flows must be clean.

Das Bild 12*b* zeigt eine Heizfläche, **welche** außen durch Ölkoks verschmutzt ist.

Figure 12*b* shows a heating surface which (*or* that) is soiled with oil coke on the outside.

Es wird ein höchster CO_2-Gehalt von etwa 13% erreicht, **dem** ein Mindest-Luftüberschuß von 30% entspricht.

A maximum CO_2 content of about 13% is reached, to which corresponds a minimum air excess of 30%.

Der Stickstoff, **dessen** Wertigkeit zwischen eins und sechs schwankt, ist hier vierwertig.

Nitrogen, whose valence fluctuates between one and six, is tetravalent here.

Warning. Not every **der, die,** or **das**-form preceded by a comma is a relative pronoun. The second condition must also be fulfilled: a dependent clause with the variable verb *last*. Even the forms **dessen, deren,** and **denen** are not always relative pronouns. (See Appendix 1.)

Die äußere Eigenform eines Kristalls kann zwar im Falle eines Kristallaggregates unentwickelt sein, der gesetzmäßige innere atomare Aufbau wird hierdurch jedoch nicht beeinflußt.

The external characteristic form of a crystal can, to be sure, be undeveloped in the case of a crystal aggregation; the regular inner atomic structure is, however, not influenced by this.

(The word **der** after the comma cannot be a relative pronoun, since the variable verb **wird** in the second clause is not *last*. The **der** is here merely a definite article and the only word that can be regarded as a connective between the two clauses is the adverb, or adverbial conjunction, **jedoch**. See Lesson 6, Sec. B, 7.)

The *relative pronoun* forms are given in Table 7.

TABLE 7. RELATIVE PRONOUNS

der, welcher	can be *masculine nominative singular* who, which, that
	or *feminine dative singular* to whom, to which
dessen and **deren**	are always *genitive case* forms and can usually be translated, whose
dem, welchem	are *masculine* and *neuter, dative singular* to whom, to which

TABLE 7 (Continued)

den	is *masculine accusative singular*	whom, which, that
denen	is *dative plural*	to whom, to which
welchen	can be equivalent to either **den** or **denen**	
die, welche	can be *plural* and *feminine singular, nominative* who, which, that	
	or *plural* and *feminine singular, accusative* whom, which, that	
das, welches	are *neuter, nominative* and *accusative singular* who(m), which, that	

Note that:

(*a*) The *same* relative pronouns are used for *persons* and *things*.

(*b*) The relative pronoun has the same *gender* and *number* as its antecedent (the word it refers to). Its *case* is determined by its function within the dependent clause that it introduces.

2. The pronoun **wer** (who) sometimes occurs as a relative pronoun in the meaning of "whoever" or "anyone who," with no antecedent.

Wer sich für die Bekanntgabe neuer Erfindungen interessiert, wird manchem Patent der letzten Jahre begegnet sein, das dieses Thema behandelt.

Whoever is (*or* anyone who is) interested in the announcement of new inventions will have encountered many a patent dealing with this topic in recent years.

3. The pronoun **was** is encountered as a relative pronoun whenever there is an antecedent that has no definite gender or number. This includes:

(*a*). The indefinite pronouns:

Alles (nichts, etwas, viel), was er behauptet, ist wahr.

Everything (nothing, anything, much) that he asserts is true.

(*b*). Indefinite superlatives:

das Beste, was er leisten konnte . . .

the best he could accomplish . . .

(c). Whole clauses:

Demgegenüber wurden im praktischen Großbetrieb Ausbeuten von 165 bis 170 g verwertbarer Produkte erzielt, **was** als ein außerordentlich günstiges Ergebnis angesehen werden muß.

In comparison, yields of 165 to 170 g of exploitable products were achieved in practical large-scale operation, (*a fact*) *which* must be regarded as an extraordinarily favorable result.

(The relative pronoun **was** refers to the whole idea expressed by the first clause rather than to any individual word in this clause. In such cases the English translation is sometimes made clearer by saying "a thing which" or "a fact which" instead of simply "which.")

4. The **wo**-words are often equivalent to a preposition plus a relative pronoun.

das Probierröhrchen, **in dem** (or **in welchem**, or **worin**) sich Wasserstoff und Luft befinden

the test tube, *in which* there are hydrogen and air

Die Erscheinung kann auch ohne Lichtemission stattfinden, **woraus** zu schließen ist, daß . . .

The phenomenon can also take place without the emission of light, *from which* (*fact*) it can be concluded that . . . (The **wo** of **woraus,** and the "which" in English, refer to the whole idea expressed by the first clause.)

B. DOUBLE VERBS: A PITFALL IN FINAL POSITION

The reader or translator who has learned that he must look to the end of a clause for a verb may still run into trouble if such a clause consists of two parts connected by a co-ordinating conjunction, usually **und.**

Wir wollen die Angaben dieses Abschnittes **zusammenstellen** und dabei auch für x einen Wert **einsetzen.**

We want to summarize the data of this section and in doing so also put in a value for x.

C. ANTICIPATED CONJUNCTIONS

Some conjunctions introducing subordinate clauses are anticipated by certain adverbs in the preceding clause, which serve as a sort of warning signal for the approaching conjunctions. Often the anticipating adverb cannot be explicitly translated.

Die Gase hatten **so lange** das Streben nach oben, **bis** sich ihre Temperatur derjenigen der Luft angeglichen hatte.

The gases had the tendency to rise (for such time) until their temperature had become equal to that of the air.

Dieser Effekt tritt **erst dann** (or **nur dann**) ein, **wenn** längliche Teilchen in der Lösung vorhanden sind.

This effect does not occur until (*or* occurs only if) elongated particles are present in the solution.

Es gibt dafür verschiedene graphische Darstellungsmöglichkeiten, die hier **deshalb** näher beschrieben werden, **weil** sie für viele ähnliche Systeme entsprechende Bedeutung haben.

There are various possibilities of graphic representation of this which are more closely described here because (**or** described . . . for the reason that) they have corresponding significance for many similar systems.

D. WORD ORDER WITHIN CLAUSES

The relative position of sentence elements other than subject and verb within any clause, dependent or independent, was described in part in Lesson 2, Section C. Other details in this category are the *extended modifier construction,* taken up in detail in Lesson 12, Section B, and the *detached verb prefixes,* presented in detail in Lesson 10, Section B. One point remains to be mentioned here:

It is characteristic of German to use a great many idiomatic adverbs, some of which represent fine shades of meaning that are practically impossible to translate. They are particularly difficult to deal with when they occur in close succession. The best translation technique is to separate them in the English sentence:

Auch heute **noch** hört man diesen Ausdruck.
One *still* hears this expression (*even*) today.

Von allen diesen Angaben gilt **also nur noch** diese eine.
Thus, of all these specifications, *only* this one is *still* valid.

EXERCISE
DIE VERWENDUNG KÜNSTLICH RADIOAKTIVER ISOTOPE IN DER MEDIZIN

Von Professor Kurt Phillip, Freiburg i. Br.

Durch die Entdeckung der künstlichen Radioaktivität besitzen wir die Möglichkeit, für fast jedes chemische Element des Periodischen Systems radioaktive Isotope zu verwenden, d.h. Atomarten verschiedener Masse, welche die gleiche Kernladung besitzen, somit[1] an die gleiche Stelle des Periodischen Systems gehören und damit auch die gleichen chemischen Eigenschaften haben. Infolge ihrer Strahlung können sie in außerordentlich geringen Mengen in einfacher Weise z.B. mit einem Zählrohr nachgewiesen werden. Man kann daher, wie *G. von Hevesy* vor ca. 30 Jahren in seinen Arbeiten über die Indikatormethode erstmalig gezeigt hat, einem Stoff radioaktive Isotope, also[2] markierte Atome, zusetzen und dadurch den Verbleib des Stoffes etwa im pflanzlichen oder tierischen Organismus besonders leicht verfolgen. Handelt es sich hierbei um die Elemente selbst, und spielt es keine Rolle, in welcher Form oder Verbindung sie vorliegen, so braucht man nur das entsprechende radioaktive Isotop direkt oder in einer Verbindung zuzusetzen. Möchte man jedoch den Verbleib einer bestimmten Verbindung untersuchen, muß man diese in geeigneter Weise durch Einbau eines radioaktiven Isotopes markieren. Dies kann bei sehr komplizierten organischen Verbindungen auf Schwierigkeiten stoßen, denen man in manchen Fällen dadurch aus dem Wege geht, daß man die Synthese dieser Stoffe durch einen pflanzlichen oder tierischen Organismus bewirken läßt. Wohl die erste solcher Synthesen ist 1941 von

Born, Lang, Schramm und *Zimmer* durchgeführt worden. Es[3] wurden Tabakpflanzen in radiophosphorhaltiger Nährlösung aufgezogen und unmittelbar nach der Radiophosphorzugabe mit dem Tabakmosaikvirus infiziert. Nach 4 Wochen zeigte das abgetrennte Virusprotein eine hohe Aktivität. Ähnlich konnten *A. Niklas* und *W. Maurer* aus der Hefe Torula auf einem mit radioaktivem Schwefel (^{35}S) versetzten Nährboden[4] Methionin mit einer außerordentlich großen spezifischen Aktivität (mehrere mC pro 1 mg Methionin) herstellen. In Amerika hat man sogar einen großen radioaktiven Garten angelegt, in welchem der Boden und die Luft und damit die darin wachsenden Pflanzen radioaktiv gemacht werden können. Man kann auf diese Weise über[5] die entsprechenden Pflanzen radioaktives Morphin (Schlafmohn), Digitalis (Fingerhut), Atropin (Belladonna), Nikotin (Tabak) oder auch das so außerordentlich interessierende Chlorophyll radioaktiv markieren und den Wirkungsmechanismus dieser Stoffe studieren.

Physikalische Blätter 11, pp. 260–261, Physik Verlag, Mosbach, 1955.

Notes for Exercise

[1] **somit,** and thus. This adverbial conjunction merely adds another part to the dependent clause beginning with **welche:** which possess . . . and (which) thus belong. . . .

[2] **also,** that is.

[3] Ignore the **Es.** The subject is **Tabakpflanzen.** See Lesson 9, Section C.

[4] **auf einem . . . Nährboden = auf einem Nährboden, der mit radioaktivem Schwefel versetzt ist.** Original wording is an Extended Modifier Construction. See Lesson 12, Section B.

[5] **über** here means "by way of."

LESSON 8

Verbs—Part One

Simple and Compound Tenses

As indicated in Lesson 3, German verb forms are either simple (consisting of one word) or compound (consisting of several forms). This closely parallels the structural pattern of English verbs.

A. SIMPLE FORMS

1. *Forms Ending in* -e

(*a*). The **ich**-form (first person singular) of the *present tense*.
    ich zeige   I show, I am showing, I do show

Combinations with the English auxiliary *do* (*does*, past: *did*) are chiefly necessary for questions (**lese ich,** do I read) and negatives (**ich lese nicht,** I do not read). It is also important to know that the German present tense, even more so than the English present, can refer to *future* time:

Ich **fahre** nächste Woche **ab.**   I leave (am leaving, will leave) next week.

(*b*). The exhortation form (always third person singular).
    man zeige   (let one) show (see item 6 below)

Simple Forms

2. Forms Ending in -t

This is the characteristic ending of the third person singular of the present tense, with the subjects **er, sie, es**:

er zeigt	he shows, he is showing, he does show (from **zeigen**)
er arbeitet	he works, he is working, he does work (from **arbeiten**)
er trägt	he carries, he is carrying, he does carry (from **tragen**)
er hält	he holds, he is holding, he does hold (from **halten**)
er gibt	he gives, he is giving, he does give (from **geben**)
er liest	he reads, he is reading, he does read (from **lesen**)

The last four verbs listed here illustrate an important principle. Verbs with certain specific series of stem changes have an internal change in the third person singular of the present tense:

tragen	trug	getragen	trägt	(a—u—a—ä)
halten	hielt	gehalten	hält	(a—ie—a—ä)
geben	gab	gegeben	gibt	(e—a—e—i)
lesen	las	gelesen	liest	(e—a—e—ie)

To these we can add:

brechen	brach	gebrochen	bricht	(e—a—o—i *or* ie)
schmelzen	schmolz	geschmolzen	schmilzt	(e—o—o—i)

The reader must learn to identify these by experience. It is particularly important to realize that the *umlauted* forms (*trägt, hält,* etc.) are *present tense* forms, in order to avoid confusion with the *short conditional,* explained below.

3. Forms ending in -en (*or* -n *after* l *and* r)

(*a*). These can be infinitives especially if preceded by *zu:*

Man versuchte **zu ermitteln.** . . .

They tried to determine. . . .

Infinitives without **zu** occur as parts of compound tenses (future and long conditional) and with non-tense-forming auxiliaries:

Man wird (würde, kann, muß, soll) es **ermitteln**.
They will (would, can, must, are supposed to) determine it.
Er läßt es **ermitteln**.
He is having it determined.

(*b*). The forms in **-n** or **-en** can be present tense plurals:

>sie **gehen**   they go, are going, do go
>sie **handeln** they act, are acting, do act

The principles of word order must be used as a guide in translating forms ending in **-en** as (*a*) infinitives, or (*b*) present tense forms agreeing with a plural subject:

Wir werden diese Dinge später **besprechen**. (*Infinitive*)
We will discuss these things later.

. . . die Dinge, die wir hier **besprechen** wollen (*Infinitive*)
the things we want to discuss here.

die Dinge, die wir hier **besprechen** (*Present*)
the things we are discussing here

Wir **besprechen** hier nur die einfachen Formen. (*Present*)
We are discussing only the simple forms here.

Warning. The forms with **-e** and **-en** endings must be distinguished from the forms with **-te** and **-ten** endings, and the forms with **-en** endings are not present tense forms unless they are identical with the infinitive of the verb.

4. *Formation of the Past Tense*

As already pointed out in Lesson 3, German verbs form their past tenses in the same two ways as English verbs, i.e.,

(*a*) by adding an ending to the stem (so-called *weak* verbs):

>fragen   fragte
>ask   asked

(b) by undergoing an internal change and adding no ending (in the singular) (so-called *strong* verbs):

>sehen   sah
>see   saw

There are also a few weak verbs that undergo an internal change besides adding the -te ending:

> **haben—hatte** **brennen—brannte***
> have—had   burn—burned
>
> **bringen—brachte** **wenden—wandte**†
> bring—brought   turn—turned
>
> **denken—dachte** **wissen—wußte**
> think—thought   know—knew

In referring to them as a group, the strong verbs and the irregular weak verbs can be called *stem-changing verbs*, while the regular weak verbs can be called *non-stem-changing*.

(*a*). Forms ending in -te (singular) and -ten (plural). These are usually past tenses of weak verbs and are easily traced to their infinitive by simply substituting -en for the -te or -ten (in some cases for -ete and -eten):

>**zeigte** → **zeigen**   **arbeitete** → **arbeiten**

But there are pitfalls because:

1. The forms, especially if they have an umlaut, may be *short conditionals*. This is particularly true of the irregular weak verbs and the non-tense-forming auxiliaries:

> **wußte** knew   **mußte** had to
> **wüßte** would know **müßte** would have to

2. The -t before the -e or -en may be part of the verb stem:

---

* Three other verbs undergo similar changes:
  nennen to mention, **rennen** to run, **kennen** to be acquainted with

† One other verb undergoes the same changes: **senden** to send. Both **senden** and **wenden** also occur as weak verbs with the past tense forms **sendete** and **wendete.**

ich arbeite	I am working	(present of **arbeiten**)
wir arbeiten	we are working	
Man beachte dieses Beispiel.	Observe this example.	(Exhortation form of **beachten**).

(*b*). Past tenses of strong verbs:

er **sah**	he saw, he was seeing, he did see*
er **ging**	he went, he was going, he did go
sie **sahen**	they saw, they were seeing, they did see
sie **gingen**	they went, they were going, they did go

The student must gradually acquire an ability to recognize such forms as past tenses. Dictionaries list the singular forms with a cross-reference to the infinitive. But in order to be at all proficient in reading, it is practically indispensable to be able to identify stem-changing verbs of all types in all of their forms, not only the past tense but also the past participle and the singular of the present tense when it involves an additional internal change.

The reader who has to stop to look up each such form is wasting valuable time and making his task unnecessarily tedious. The best way to know the forms is to memorize the principal parts of the most frequent verbs. A fairly complete list of these is given in Appendix 6, and the alphabetical list of verb forms in Appendix 7 may be a valuable guide in determining which verbs are the most frequent.

5. *Short Conditional Forms*

The past tenses of stem-changing verbs having a stem vowel **a, o** or **u** may appear with an umlaut on this vowel. The singular form in the case of strong verbs then ends in **-e**. Past tenses with other stem vowels appear modified only by the **-e** ending in the singular.

---

* These are the 3 possible interpretations of *all* simple past tenses. Cf. Item 1 (*a*).

Simple Forms

wäre, wären	would be	fiele	would fall
wüßte, wüßten	would know	ginge	would go
müßte, müßten	would have to		
hätte, hätten	would have		
		würde	would
gäbe, gäben	would give		
flösse, flössen	would flow		

The forms *brennte, kennte, nennte, rennte* serve as short conditionals for the verbs *brennen, kennen, nennen, rennen* respectively, while *sendete* and *wendete* are the corresponding forms for *senden* and *wenden*.

The short conditional is particularly frequent for auxiliary verbs in complex combinations:

hätte gefragt	*would have* asked
dürfte anfangen	*would be allowed* to start
hätte schreiben können	*would have* been able to write
	or could have written
könnte schreiben	*could* (= *would be able to*) write

In *if*-clauses, *as-if*-clauses, and sometimes in indirect quotations, it is best to render the short conditional with an English past tense:

Wenn wir das **wüßten**. . . .
If we *knew* that. . . . (meaning: if we *were to know* that)

Als ob (*or* wie wenn) es wichtig **wäre**.
As if it *were** important.

Sie behaupteten, es **käme** selten **vor**.
They claimed that it rarely *happened*.

Occasionally a past tense of a regular non-stem-changing verb is used in a conditional sense:

Wenn wir uns auf diese Ergebnisse verlassen könnten, **brauchten** wir nicht weiterzuforschen.

If we could rely on these results we *would* not *need* to investigate further.

---

* This is the only English verb that has its own distinctive form for this usage.

6. Exhortation Forms

(a). These occur in such expressions as:

Er **ruhe** in Frieden.   So **sei** es.
May he *rest* in peace.   So *be* it, *let* it *be* so.
Möge es gelingen.   **Sei** es noch so schwer.
May it *be* successful.   *Be* it (or *may* it *be*) ever so difficult.

They are always third person singular and end in **-e**, except for the verb **sein**, which has a singular **sei** and a plural **seien**:

x und y **seien** Zahlen. . . .
Let x and y *be* numbers. . . .

(b). In scientific writing they occur mainly in a few typical phrasings:

Im Vorübergehen **sei** hier noch erwähnt. . . .
In passing *let* it also *be* mentioned here (or *let us* also mention here). . . .

Es **gehe** ein Lichtstrahl zur Zeit t_A von A nach B **ab**.
Let a light ray *depart* from A toward B at the time t_A.

Man denke an die Möglichkeiten.
Think of the possibilities.

(*Note* that the exhortation form with the subject *man* is best rendered into English as an *imperative*.)

(c). The exhortation form is also extensively used for *indirect quotation*. The English translation is usually a present or a past tense, depending on the tense of the verb introducing the quotation:

Er **behauptet(e)**, es **liege** eine Legierung **vor**.
Present: He *claims* that an alloy *is* present.
Past: He *claimed* that an alloy *was* present.

The use of the exhortation form in an indirect quotation usually casts doubt on the truth of the quotation or at least disclaims all responsibility for its truth. Whether or not an author uses the exhortation form therefore depends on the extent to which he is anxious to bring out this implication and not so much on any grammatical rule. The translator may, in fact, sometimes find it necessary to rephrase the English in order to achieve the desired effect.

The *exhortation form* and the *short conditional* are both parts of what is traditionally called the *subjunctive*. This last term has been avoided here in order to spare the reader the intricacies of terminology that would arise from its use, particularly since there are two schools of thought on the subject, which use the same "tense" names with different meanings. The exhortation form and the short conditional actually have very little in common except that (*a*) both are used in indirect quotations and (*b*) both can be used in *as-if*-clauses:

 . . . , **als ob das sehr wichtig wäre** (or **wichtig sei**)
 . . . , **als wäre** (or **als sei**) **das sehr wichtig**
 as if that were very important

B. COMPOUND TENSES

The most essential information on these forms was given in Lesson 3, Section D. Only a few details need be added here:

1. The Long Conditional

The *umlauted* past tense of **werden**, accompanied by the *infinitive* of another verb, forms the *long conditional* of that verb, that is, **würde(n)** is equivalent to English "would":

Ich, er, sie, es, man **würde sein.** I (he, she, etc.) *would be.*
wir, Sie, sie **würden sein.**

This is identical in meaning to the short conditional explained above, i.e., *ich würde sein = ich wäre*.

While the auxiliary **würde** can usually be relied on to correspond to English "would," it is occasionally used in subordinate clauses as a short form for **würde werden** (would become):

..., weil die Sache dadurch nur viel schwieriger **würde**.
..., because as a result of this the matter *would* only *become* much more difficult.

2. Perfect Tenses

The auxiliaries **haben** or **sein,** in combination with the *past participle* of another verb, form the perfect tenses of that verb. They denote action completed at a time corresponding to the tense of the auxiliary:

Er **hat** Schlüsse gezogen.	He *has* drawn conclusions.	*Present Perfect* (action now completed)
Er **ist** aufs Land gezogen.	He *has* moved to the country.	
Er **hatte** Schlüsse gezogen.	He *had* drawn conclusions.	*Past Perfect* (action completed at some past time)
Er **war** aufs Land gezogen.	He *had* moved to the country.	
Er **wird** Schlüsse gezogen **haben.**	He *will have* drawn conclusions.	*Future Perfect* (action completed by some future time)
Er **wird** aufs Land gezogen **sein.**	He *will have* moved to the country.	
Er **würde** Schlüsse gezogen **haben.** (Long form) / Er **hätte** Schlüsse gezogen. (Short form)	He would have drawn conclusions.	*Conditional Perfect* (action completed if conditions had permitted)
Er **würde** aufs Land gezogen **sein.** (Long form) / Er **wäre** aufs Land gezogen. (Short form)	He would have moved to the country.	

Note that the same verb (**ziehen**) is used here with different auxiliaries and different meanings. Ordinarily all transitive and reflexive verbs form their perfects with the auxiliary **haben,** and *so do most intransitive verbs.* Only a *small number* of intransitive verbs form their perfects with **sein.** This is of some significance in consulting the dictionary; verbs with **haben** can be *v.t., v.r.,* and *v.i.* Verbs with

sein are always *v.i.*, and the dictionary often indicates that this auxiliary is used by means of some such notation as *v.i.*(s).

Warning. Not every combination of **sein** + *past participle* is a perfect tense. The past participle may also be in use as a predicate adjective and will then not be identified as *v.i.*(s) in the dictionary.

   Es ist erreicht!
   It *is* accomplished!
   (**erreichen** is a *v.t.*).

3. *The Passive Voice*

This is taken up in greater detail in Lesson 9, Section B. It is formed with the auxiliary **werden** in all tenses.

EXERCISE
EINFLUß DER REIBUNG BEI UMSTRÖMTEN KÖRPERN

Grundsätzliches zum Widerstandsproblem

Die bei der Rohrströmung gefundenen Gesetzmäßigkeiten[1] können leicht auf die Umströmung von Körpern übertragen werden. Nach Fig. 3 werde z.B. ein Zylinder in der unendlich breiten Parallelströmung betrachtet. In einer hinreichend weiten Entfernung vor dem Zylinder nehmen wir eine

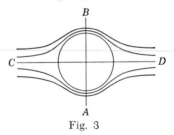

Fig. 3

gleichmäßige Einteilung vor. Das Schicksal der Einteilungsgrenzen verfolgen wir und erhalten so das Stromlinienbild. Bei reibungsfreier Flüssigkeit entsteht das bereits bekannte Bild nach Fig. 3.

Da Reibungslosigkeit vorausgesetzt ist, können wir die Begrenzung einer Stromlinie auch durch reibungsfreie Wände ersetzen. Bei der den Zylinder berührenden Stromlinie[1] sei dieses Experiment durchgeführt. Nunmehr haben wir bereits eine Rohrströmung, auf die wir die früheren Gesetze anwenden können. Dort, wo der Kanalquerschnitt eng ist, z.B. direkt über dem Zylinder, herrscht große Geschwindigkeit d.h. nach *Bernoulli* kleiner Druck. Vor dem Zylinder ist der Querschnitt am größten, was dem größten statischen Druck entspricht. Die vollkommene Symmetrie in bezug auf die Achse A-B zeigt, daß keine resultierende Kraft in Strömungsrichtung vorhanden ist, da die Drücke auf der Vorderseite des Zylinders genau denen[2] auf der Rückseite entsprechen und sich aufheben. Denkt man sich jedoch den Zylinder durch C—.—D in zwei Hälften geteilt, so würden diese nach oben und unten gesaugt werden (s.[3] auch Berechnung auf S.[3] 31). Die nach der *Bernoullischen* Gleichung berechenbare Saugwirkung[1] an den Grenzen des Zylinders nimmt infolge der Zentrifugalkräfte normal zu der Stromlinie ab. Auf S. 15 war die Druckzunahme in der zur Strömung senkrechten Richtung[1] berechnet worden, die durch die Zentrifugalkraft bedingt ist. Eine kleine Rechnung soll zeigen, wie die Drücke sich zahlenmäßig ändern.

Technische Strömungslehre, von Bruno Eck, 1. Bd., S. 80, Berlin, Springer-Verlag, 1935.

Notes for Exercise

[1] These are extended modifier constructions. See Lesson 12, Section B.
[2] **denen**, to those. See Appendix 1.
[3] s. = siehe, see; S. = Seite. page, p.

LESSON 9

Verbs—Part Two

Reflexive—Passive—Impersonal

A. REFLEXIVE AND RECIPROCAL VERBS

1. In the strict sense a *reflexive verb* is one whose direct object is a reflexive pronoun, i.e., an object pronoun referring to the same person or thing as the subject:

beherrschen to control

ich beherrsche mich	I control myself
er beherrscht sich	he controls himself
sie beherrscht sich	she controls herself
es beherrscht sich	it controls itself
man beherrscht sich	one controls oneself
wir beherrschen uns	we control ourselves
Sie beherrschen sich	you control yourself *or* yourselves
sie beherrschen sich	they control themselves

Since the reflexive pronoun is in fact an object pronoun, its position in clauses is that of object pronouns:

In dieser Abhandlung befaßt er **sich** (befaßt **sich** der Verfasser) nur mit einfachen Fällen.

In this article he concerns himself (the author concerns himself) only with simple cases.

Diese Abhandlung, in der **sich** der Autor nur mit den einfachsten Beispielen befaßt, ...

This article, in which the author concerns himself only with the simplest examples, ...

Reflexive pronouns can also serve as indirect objects or in other functions calling for a dative case. The forms **sich** and **uns** can be either dative or accusative, but **mich** is accusative only. The corresponding dative form is **mir**.

Ich erkläre mir die Sache folgendermaßen.

I explain the matter (to myself) as follows.

Sehen wir uns noch ein paar Beispiele an.

Let us look at a few more examples (for ourselves).

Note. This sort of dative reflexive pronoun can often be omitted in English.

Note. Verbs with such reflexive indirect objects are technically transitive verbs and most dictionaries list such uses under *v.t.* rather than *v.r.*

In a few verb combinations a dative reflexive pronoun adds a special meaning:

denken to think   **vorstellen** to present, introduce
sich etwas denken, sich etwas vorstellen   to imagine a thing, to conceive of a thing

Man kann es sich in der Form eines Zylinders denken.

One can imagine it (*or* picture it, conceive of it) in the form of a cylinder.

Man stelle sich vor, wie unmöglich das wäre!

Imagine how impossible that would be!

Most German reflexive verbs cannot be translated as literally as **sich beherrschen** (to control oneself) or **sich mit etwas befassen** (to concern oneself with something). The usual English translation is either a phrase or an intransitive verb:

    sich verkleinern   to grow smaller, to shrink in size
    sich erhöhen   to rise, to increase

Where an English verb can be both transitive and intransitive, the German equivalent of the intransitive use usually appears as a reflexive:

German *v.t.*	Wir **bewegen** es hin und her.	
English *v.t.*	We *move* it back and forth.	
German *v.r.*	Es **bewegt sich** hin und her.	
English *v.i.*	It *moves* back and forth.	

2. The reflexive pronoun is sometimes used in a *reciprocal* meaning, provided the context makes this interpretation clear. Otherwise a special reciprocal pronoun **einander** is used:

Sie sind **sich** gleich.
Sie sind **einander** gleich.
} They are equal *to each other* (dative).

Die Linien **schneiden sich** in A. The lines intersect at A (accusative).

Note that the English translation need not always express the reciprocal pronoun *each other*. This is also true of prepositional compounds with **einander**:

Die Substanzen reagieren nicht **miteinander**.
The substances do not react (with each other).

B. PASSIVE VOICE

1. With Subject Expressed

The passive voice constitutes a series of compound tenses, in which the verb appears as a past participle accompanied by the appropriate tense of an auxiliary verb: in English *to get** or *to be*, in German **werden** (to get, to become).

Present
  Das Buch **wird** von allen Forschern **gelesen**.
  The book *is read* by all scientists.

---

* As in the phrasing "*got* caught" for "*was* caught," German: **wurde** erwischt.

Past

    Das Buch **wurde** von allen Forschern **gelesen**.
    The book *was read* by all scientists.

Future

    Das Buch **wird** von allen Forschern **gelesen werden**.
    The book *will be read* by all scientists.

Conditional

    Das Buch **würde** von allen Forschern **gelesen werden**.
    The book *would be read* by all scientists.

Present Perfect

    Das Buch **ist** von allen Forschern **gelesen worden**.
    The book *has been* (or *was*) *read* by all scientists.

Past Perfect

    Das Buch **war** von allen Forschern **gelesen worden**.
    The book *had been read* by all scientists.

Future Perfect

    Das Buch **wird** von allen Forschern **gelesen worden sein**.
    The book *will have been read* by all scientists.

Conditional Perfect (short form)

    Das Buch **wäre** von allen Forschern **gelesen worden**.
    The book *would have been read* by all scientists.

Note that **worden** occurs in all of the perfects, and only in the perfects. It is equivalent to English "been."

In subordinate clauses:

    . . . ein Buch, das von allen Forschern **gelesen wird** (**gelesen wurde, gelesen werden wird,** etc.)
    . . . a book that *is* (*was, will be,* etc.) *read* by all scientists.

With modal auxiliaries:

Das Buch **kann** (**muß, soll, sollte**) von allen Forschern **gelesen werden**.

    The book *can* (*must, is supposed to, should*) *be read* by all scientists.

    . . . ein Buch, das von allen Forschern **gelesen werden sollte**.
    . . . a book that *should be read* by all scientists.

The form **würde** sometimes appears with a past participle; this is short for **würde werden**:

... wenn das Buch auch von allen Forschern **gelesen würde**.
... even if the book *were* (*literally: would be*) *read* by all scientists.

The present (**wird gelesen**, "is read") is by far the most frequent passive form, but great care must be taken not to confuse it with an *active future*. This is an especially frequent error where a stem-changing verb is involved whose past participle is identical with its infinitive:

halten   hielt   gehalten   hält   to hold
erhalten   erhielt   **erhalten**   erhält   to obtain

Only the context can help to decide whether a combination like **wird erhalten** is future (*will obtain*) or present passive (*is obtained*):

... ein Resultat, das nur von einem geübten Fachmann erhalten wird.
... a result that is obtained only by a practiced expert.

... ein Resultat, das nur ein geübter Fachmann erhalten wird.
... a result that only a practiced expert will obtain.

2. *Subjectless Passive*

In English the passive voice is practically limited to transitive verbs. In German, the emphasis on an action going on, without reference to its performer, is applied to intransitive verbs as well. The result is a "subjectless" passive, which is very difficult to translate. It is often equivalent to the use of the pronoun *man*.

Man experimentierte jahrelang.
Es wurde jahrelang experimentiert.

They experimented for years, *or*
Experimenting was done for years.

The word **es** used at the beginning of the passive sentence is not a subject. It is merely a sentence-beginning element to keep the verb in second position. Any other sentence element can take its place, in which case the es is dispensed with entirely:

In Amerika wurde jahrelang experimentiert.
In America experimenting was done for years.

The English language does have a device that can often be applied to the translation of a subjectless passive if it is accompanied by a prepositional phrase. This device consists of putting the preposition somewhere after the verb and using the object of the preposition as an apparent subject:

Mit solchen Apparaten wurde in Amerika jahrelang experimentiert.
Such devices were experimented *with* for years in America.
Auf diese Einzelheiten wird erst später eingegangen.
These details will not be gone *into* until later.

C. NONPERSONAL AND IMPERSONAL EXPRESSIONS

1. Certain verbs in German can never have a *person* as their subject. This is troublesome in a number of situations where the same idea has to be expressed *with a personal subject* in English. The outstanding example is that of the verb expressing the idea of being successful:

gelingen   gelang   ist gelungen

Basically this verb means "to be possible," "to be successful."
Der Versuch gelang.
The experiment was successful.
Es gelang, alle diese Probleme zu lösen.
It was (*or* proved) possible to solve all these problems.
All these problems were successfully solved.

To indicate that a *person* is successful, the person is introduced in the dative case:

Es gelang **dem Forscher,** alle diese Probleme zu lösen.
It was possible *for the research man* to solve all these problems.
The research man succeeded in solving all these problems.

Der Versuch gelang **ihm.**
The experiment was successful for him.
He succeeded in (*or* with) the experiment.

2. Certain verbs in German are characterized by the fact that they occur only with the subject **es**, at least in certain meanings. This generally corresponds to the use of *it* as a rather meaningless and vague subject for expressions of the weather

                Es regnet   It is raining
                Es schneit   It is snowing

and predicate adjectives introducing subordinate or infinitive clauses:

Es ist ratsam, ein paar Minuten zu warten.
It is advisable to wait a few minutes.

But sometimes the translation is more difficult:

Es donnert und blitzt. There is thunder and lightning.

Es kracht im Gebälk. There is a creaking (noise) in the woodwork.

Certain idiomatic expressions are particularly frequent:

Es handelt sich um radioaktive Isotopen.
It is a matter of radioactive isotopes.

Es gibt auch kompliziertere Kombinationen.
There are also more complicated combinations (stresses *existence*).

Er behauptete, daß **es** dort fast keinen atmosphärischen Druck **gebe.**
He maintained that *there was* almost no atmospheric pressure there.

Hier **kommt es** hauptsächlich **darauf an,** diese Folgerungen über die Natur des Mesons zu prüfen.
Here *it is* mainly *a matter of* testing (*or it is* mainly *important to* test) these inferences concerning the nature of the meson.

Note that in all of the above examples, the **es** is an essential part of the expression and is always present regardless of word order.

3. There are situations where it is expedient to put the subject of a clause closer to the middle or end than it would ordinarily stand:

A *time* comes in everyone's life, when . . .

There comes *a time* in everyone's life, when . . .

Note that in order to delay the real subject "a time" we insert the word "there" as a sort of filler, since we cannot leave the space before the verb completely blank. In German this function is performed by the pronoun **es**, with the following provisions:

(*a*). The word **es** must begin the sentence. If anything else stands first, **es** is omitted because the subject must then follow the verb anyway:

Es kommt im Leben eines jeden ein Augenblick, wo . . .

Im Leben eines jeden kommt ein Augenblick, wo . . .

(*b*). The verb following **es** may be singular or plural, depending on the real subject:

Es ist ein Brief da.	There is a letter here.
Es sind Briefe da.	There are letters here.

(*c*). The principle of delaying the subject and starting with a filler-word (**es** in German, *there* in English) is applied far more extensively in German than in English, hence cannot usually be translated literally. The technique is to find the real subject and start with that:

Es sind radioaktive Isotope vorhanden.

There are radioactive isotopes present.

Radioactive isotopes are present.

Es blieb nichts übrig.

Nothing remained (*or* was left over).

Es lassen sich noch keine genauen Regeln aufstellen.

No exact rules can as yet be set up.

Es würde dies bedeuten, daß . . .

This would mean that . . .

Note that if the verb is *plural,* there can be no doubt that you are dealing with such a construction. If the verb is *singular,* then the initial **es** may just as well be an actual pronoun referring to a previously mentioned noun, or it may be in use as the impersonal subject discussed in item 2. above.

For the use of the expletive **es** with the subjectless passive see Section B, 2. of this lesson.

EXERCISE
WANN GILT DAS STABILITÄTSKRITERIUM NACH NYQUIST?

Von Johannes Peters, Hamburg

Es wird ein allgemeines Gleichungssystem für aktive Netzwerke aufgestellt, bei dem zwischen einem passiven und einem aktiven Anteil unterschieden wird. Um praktischen Gesichtspunkten zu entsprechen, wird auch der passive Anteil noch in zwei verschiedene Netzwerke aufgeteilt. Alle drei Teil-Netzwerke besitzen ein gemeinsames Knotensystem, sind also parallelgeschaltet, wobei aber nicht vorausgesetzt wird, dass jedes Teilnetzwerk bereits für sich ein zusammenhängendes Ganzes bildet.

Als Kriterium für die Stabilität des Netzwerkes wird das bekannte, auf den Wurzeln der Nennerdeterminante beruhende Kriterium benutzt und nach einem Verfahren der Funktionentheorie mit dem Stabilitätskriterium nach *Nyquist* verglichen.

I. Definitionen und Bezeichnungen in einem aktiven Netzwerk

Ein lineares passives Netzwerk besitze $n + 1$ Knoten. Dann sind n selbständige Außenstromkreise möglich. Das Netzwerk schafft lineare Beziehungen zwischen n Spannungen und n Strömen. Sind alle n Ströme bekannt, so ergeben sich die n Spannungen auf Grund der linearen Abhängigkeit und umgekehrt. Alle elektrischen Größen sind auch dann bestimmt, wenn insgesamt n Größen bekannt sind, deren Anzahl sich beliebig auf Ströme und Spannungen verteilen kann.

Eine Verfügung über Ströme oder Spannungen liegt z.B. dann vor, wenn ein unabhängiger Stromkreis kurzgeschlossen oder geöffnet wird. Dann verschwindet die entsprechende Spannung bzw. der entsprechende Strom. Die Zahl der unabhängigen Größen vermindert sich mit jedem beseitigten Freiheitsgrad um 1. Die Anschaltung eines passiven Außenstromkreises bedeutet die Beseitigung eines Freiheitsgrades. Koppelt man zwei Knotenpaare miteinander über einen passiven Vierpol im Außenstromkreis, so verschwinden drei Freiheitsgrade.

Soweit diese Änderungen der Netzwerkbeziehungen durch Außenstromkreise bewirkt werden, welche aus passiven Elementen bestehen, so kann man diese einfach dem bisherigen passiven Netzwerk hinzurechnen, wobei dann der entsprechende Außenstromkreis im neuen Netzwerk fortfällt.

Eine Einschränkung der Anzahl an Freiheitsgraden tritt auch dann ein, wenn Knotenpaare über Verstärker miteinander verbunden werden. Man kann diese Verstärker zunächst auch als eine nicht zum Netzwerk gehörige Kopplung ansehen. Ebensogut kann man aber von einem besonderen Verstärkernetzwerk sprechen, welches mit dem passiven Netzwerk gemeinsame Knotenpunkte besitzt und daher einen Bestandteil des[1] zu betrachtenden allgemeinen aktiven Netzwerkes[1] darstellt.

Archiv der elektrischen Übertragung, Bd. IV, S. 17. Wiesbaden-Biebrich, 1950.

---

[1] of the universal active network to be considered. Cf. Lesson 12 B, "Future Participle."

LESSON 10

Verbs with Prefixes

A. INSEPARABLE

The unstressed prefixes
          be-, emp-, ent-, er-, ge-, ver-, zer-
are permanently attached to verbs to create new meanings. Verbs with such prefixes have no **ge-** in the past participle (those having a **ge-** prefix to begin with naturally retain this **ge-** for the past participle).

bekommen	bekam	bekommen		(to get, receive)
empfangen	empfing	empfangen	empfängt	(to receive)
entstehen	entstand	ist entstanden		(to arise)
ergeben	ergab	ergeben	ergibt	(to yield)
gehören	gehörte	gehört		(to belong)
verändern	veränderte	verändert		(to change)
zerfallen	zerfiel	ist zerfallen	zerfällt	(to disintegrate)

Only two of these prefixes have reliable meanings that would help the reader interpret such verbs.

Ent- often implies removal or escape, hence has a somewhat negative significance:

entkommen	to get away, escape
entladen	to discharge
entkaffeinieren	to decaffeinate

Verbs with the prefix **ent-** are often used with a dative case:

Wir entkamen **der** Gefahr.
We escaped the danger.

Das Natrium zersetzt das Wasser an der Kathode und entnimmt **ihm** ein Hydroxylradikal.
The sodium decomposes the water at the cathode and takes *from it* a hydroxyl radical.

The prefix **zer-** usually implies something going to pieces or disintegrating:

zerfallen	to fall apart, decompose, disintegrate
zerbrechen	to break to pieces, shatter
zertrümmern	to shatter, smash
zerstören	to destroy

The prefixes **be-**, **er-** and **ver-** are often used to make verbs from adjective or noun stems. These verbs usually indicate bringing about the condition described by the adjective or noun:

fest	solid, firm	→	**befestigen**	to attach, make firm
Mangel	defect	→	**bemängeln**	to find fault with
Mantel	coat, cloak	→	**bemänteln**	to cloak, disguise
Anstand	objection	→	**beanstanden**	to object to
hoch	high	→	**erhöhen**	to raise
niedrig	low	→	**erniedrigen**	to lower
warm	warm	→	**erwärmen**	to heat
weit(er)	wide(r)	→	**erweitern**	to widen
möglich	possible	→	**ermöglichen**	to make possible
Gold	gold	→	**vergolden**	to gild, to gold-plate
Nickel	nickel	→	**vernickeln**	to nickel-plate
nicht	not	→	**vernichten**	to annihilate
wirklich	real	→	**verwirklichen**	to realize, bring into being
Ursache	cause	→	**verursachen**	to cause

B. SEPARABLE PREFIXES

1. These are independent words attached to verbs with more or less literal and specific meanings, although the resulting verb may then take on additional figurative meanings, e.g.,

auf up + **nehmen** to take = **aufnehmen** to take up
Feuchtigkeit aufnehmen to take up (*or* to absorb) moisture

The same verb, by an interesting extension of the meaning *to absorb*, also means *to photograph* and *to record!*

Sometimes only a figurative meaning now exists, and it is practically impossible to derive the meaning of the combination from its parts:

fangen	to catch	**anfangen**	to begin, start
hören	to hear	**aufhören**	to stop, cease

Separable verbs are basically comparable to English combinations like *take up, put out, throw away, hang fire, cut short*. Such combinations are essentially units in meaning, yet can be separated (*"Throw* those old clothes *away"*) and can have both literal and figurative meanings (consider: take off, put out, take in, make up, etc.).

Separable prefixes are always stressed.

Separable prefixes appear in two possible locations in the sentence:

(*a*). *Detached.* Whenever the verb to which they belong is in a simple tense and located in first or second position.

A detached prefix is always at the end of its clause:

Man kommt (kam, käme) immer wieder auf dasselbe Prinzip **zurück.**

We come (came, would come) back again and again to the same principle.

(*b*). *Attached.* To any verb form in last or next to last position. This includes all constant forms, and any variable forms that come last in dependent clauses.

Wir werden (würden, müssen, können, sollten) immer wieder auf dasselbe Prinzip **zurückkommen.**
We will (would, must, can, should) come back again and again to the same principle.

Wir versuchen, immer wieder auf dasselbe Prinzip **zurückzukommen.**
We try to come back again and again to the same principle.

Wir sind (waren, wären) immer wieder auf dasselbe Prinzip **zurückgekommen.**
We have (had, would have) come back again and again to the same principle.

Die Diskussion wurde wieder auf dasselbe Prinzip **zurückgeführt.**
The discussion was led back to the same principle again.

Die Diskussion ist auf dasselbe Prinzip **zurückzuführen.**
The discussion is to be led back to the same principle.

Wenn man wieder auf dasselbe Prinzip **zurückkommt (zurückkäme),** ...
If one comes (were to come) back to the same principle again, ...

Wenn wir wieder auf dasselbe Prinzip zurückkommen müssen (zurückgekommen sind, zurückzukommen versuchen), ...
If we must come (have come, try to come) back again to the same principle, ...

Dictionaries use various notations to show that a verb is separable. All of them, especially the hyphen, are makeshifts and artificial; the only spelling used in actual practice is the spelling as one word. Thus the verb *aufnehmen* may appear in dictionaries as **auf-nehmen, auf/nehmen** or **auf=nehmen.**

The infinitive of one verb may serve as a prefix for another: **stehenbleiben** to remain standing, stand still, stop (moving):

Plötzlich blieb alles stehen. } Suddenly everything
Plötzlich ist alles stehengeblieben. } stopped.

Occasionally a separable prefix is attached to a verb that already has an inseparable one:

anerkennen past participle: **anerkannt** zu-infinitive: **anzuerkennen**

Wir erkennen die neue Regierung jenes Landes nicht an.
We do not recognize the new government of that country.

2. The number of separable prefixes is theoretically unlimited. Some occur on only one or two verbs (e.g., **teil** in **teilnehmen, statt** in **stattfinden** and **statthaben**), others can be attached at will to any verb where they make sense, e.g., **zurück** (back): **zurückgeben,** to give back; **zurückgehen,** to go back; **zurücklaufen,** to run back; **zurückeilen,** to hurry back; **zurückstehen,** to stand back; etc. In practice the only final arbiter of whether a particular word is to be regarded as an attachable (hence also detachable) prefix is the usage of the majority of recognized German writers as reflected in the latest revision of the standard manual of German spelling, *Duden: Rechtschreibung*. Only the whims of usage can be held responsible for such inconsistencies as **haushalten** (to keep house) vs. **Platz nehmen** to take a seat, sit down) or **zurechtmachen** (to set straight, to prepare) vs. **zunichte machen** (to destroy, ruin) and **zutage treten** (to come to light). In the final analysis **zunichte, zutage** and many other such adverbs (**zugute, zustande, imstande, zustatten, abhanden,** etc.) are just as closely associated with their verbs as are the ones that are actually attached in writing, and what is more, *their position in a sentence is invariably that of a separable prefix:*

Dieser neue Befund machte seine Theorie **zunichte.**
This new finding destroyed his theory.

Dieser neue Befund hat seine Theorie **zunichte** gemacht.
Dieser neue Befund wird seine Theorie **zunichte** machen.
. . . ein Befund, der seine Theorie **zunichte** macht.
. . . um seine Theorie **zunichte** zu machen.

Dictionaries list such combinations under the adverb (*or*

adjective *or* noun, etc.) which sometimes has no independent meaning of its own (e.g., **zustande, zutage, abhanden**):

zustande: zustande bringen, to accomplish; **zustande kommen,** to come to be, to be accomplished

gerecht *adj.*, just, fair; **gerecht werden** (+ *dat.*), to conform (to)

Platz *m* (-es, ⁔e) *1.* room, space. *2.* place. *3.* seat: **Platz nehmen,** to take a seat, to sit down

Combinations of a verb and a noun object forming a single idea, such as **Platz nehmen,** are by no means uncommon. The translation is often a single English word, or the passive voice of a verb:

Bezug nehmen (auf)	to refer, to have reference (to)
Verwendung finden	to find use, to be used
Geltung haben	to have validity, to be valid
Ausdruck finden	to be expressed

Heutzutage **findet** das Wasserstoffsuperoxyd beim Raketenbau **Verwendung.**

Nowadays hydrogen peroxide *is used* in rocket construction.

3. Even *whole prepositional phrases* may be used as if they were separable prefixes, for example:

in Anspruch nehmen	to take up, claim (e.g., *time, attention,* etc.)
in Angriff nehmen	to attack, take up (e.g., *a problem, work*)
zum Ausdruck bringen	to bring out, express
zum Ausdruck kommen	to come out, be expressed
in Betracht kommen	to come into consideration
in Betracht ziehen	to take into consideration
in Frage kommen	to come into consideration, be involved
in die Höhe gehen	to go up, rise
zur Rede kommen	to come up for discussion, be discussed
zur Verfügung stehen	(+ *dat.*) to be at the disposal (of), be available (to)
zur Verfügung stellen	(+ *dat.*) to put at the disposal (of), make available (to)

zum Vorschein kommen to appear, make its appearance
mit sich bringen to bring with it, involve
vor sich gehen to proceed

Ein solches Unternehmen nimmt (nahm, nähme) zuviel Zeit **in Anspruch.**

Such a project takes up (took up, would take up) too much time.

Ein solches Unternehmen wird (würde, kann, könnte, dürfte) zuviel Zeit **in Anspruch** nehmen.

Such a project will (would, can, could, might) take up too much time.

Ein solches Unternehmen droht, zuviel Zeit **in Anspruch** zu nehmen.

Such a project threatens to take up too much time.

Da ein solches Unternehmen sehr viel Zeit **in Anspruch** nimmt, ...

Since such a project takes up very much time . . .

Note how the phrase **in Anspruch** takes exactly the same position in all the above examples as would a separable prefix.

Verb combinations involving such phrases are listed by dictionaries under the noun in the phrase, i.e., **in Anspruch nehmen** would be found in the entry **Anspruch.** When it comes to combinations like **mit sich bringen,** practice varies; one might find this listed under **mit** or under **bringen.**

C. AMBIVALENT PREFIXES

Just as in English the word *under* enters into combination with the verb *go* both as a detached element (*to go under*) and as a permanently attached one (*to undergo*), so in German there are a few prefixes that can be either separable or inseparable. Fortunately, it is a rare thing to have two verbs made up of the same elements and used with equal frequency. The verb **übersetzen** (inseparable) means *to translate;* the verb **über/setzen** (separable) means *to ferry across*, but obviously the latter meaning is not likely to appear with any frequency in technical and scientific writing:

Der Aufsatz wurde von einem prominenten Physiker übersetzt.
The essay was translated by a prominent physicist.

Die Truppen wurden möglichst schnell übergesetzt.
The troops were ferried across as quickly as possible.

The most frequent dual prefixes are:

	INSEPARABLE		SEPARABLE	
durch	durchsehen	to see through (e.g., *a scheme*)	durch/sehen	to look through, examine
über	übergehen	to overlook, skip	über/gehen	to pass over, change
um	umgehen	to circumvent	um/gehen (mit)	to go about (with)
unter	unterhalten	to entertain	unter/halten	to hold under
wieder	wiederholen	to repeat	wieder/holen	to fetch back

The above five prefixes are likely to occur with more or less equal frequency as separable or inseparable prefixes, though not on the same verbs. There are some additional dual prefixes which, however, are in the vast majority of occurrences, especially in scientific writing, *inseparable:*

hinter	hinterlassen	to leave behind
voll	sich vollziehen	to take place
wider	widerstehen	to resist

For the reader the dual prefixes need pose no special problems not already encountered with the exclusively separable or inseparable ones. The verb is always easy to identify if the prefix is attached, and if the prefix appears at the end of the clause it is obviously separable. Dictionaries list such pairs as **übergehen** (to skip) and **über/gehen** (to pass over) as two separate entries. Unless the reader can tell from the form he encounters whether the verb is separable or inseparable, he would do well to look at *both* entries.

To one who has never seen the verb before, the forms **übergehen** or **übergeht** at the end of a clause give no clue as to whether the verb is separable or inseparable, but **übergegangen und überzugehen** are obviously separable, while **übergangen und zu übergehen** are just as obviously inseparable.

E. REDUNDANCY WITH SEPARABLE PREFIXES

Das Licht strahlt **durch** das Prisma **hindurch.**
The light shines through the prism.

In German there seem to be *two* words meaning "through" (one a preposition, the other a prefix). This is a frequent occurrence and does not seem repetitious to Germans. In English, of course, only one of the two words, the preposition, can be expressed.

EXERCISE
PHOTOSYNTHESE

Von Otto Warburg

Aus einem Vortrag 1948

Belichtet man grüne Pflanzenzellen, die in Berührung mit kohlensäurehaltiger Luft sind, so wird Kohlensäure unter Entwicklung von Sauerstoff zu Zucker reduziert. Durch diese Reaktion, die „Photosynthese", wird in der lebenden Natur die Energie des Sonnenlichts in chemische Arbeit verwandelt. Es ist eine Reaktion, von der wir alle leben, und doch war ihr Mechanismus bis vor wenigen Jahren ein Geheimnis der lebenden Natur. Heute nähern wir uns der Lösung des Problems.

Die Pflanzenzellen, in denen sich die Photosynthese abspielt, enthalten grüne Organe, die Chloroplasten, die ihrerseits sehr kleine grüne Körnchen, die grünen Granula, enthalten. Das gesamte Chlorophyll der Pflanzenzellen ist in den grünen Granula konzentriert und so sind es die grünen Granula, in denen das bei der Photosynthese wirksame Licht absorbiert wird.

Sowohl die Chloroplasten als auch die grünen Granula kann man aus den Pflanzenzellen leicht isolieren. Aber weder die isolierten grünen Granula noch die isolierten Chloroplasten sind der Photosynthese fähig, sondern zur Photosynthese ist die gesamte intakte Pflanzenzelle notwendig. Drückt man ein Blatt zwischen zwei Glasplatten, bis die Konturen der Chloroplasten innerhalb der Zellen unscharf geworden sind, so ist die Fähigkeit der Zellen zur Photosynthese vollständig und unwiederbringlich verschwunden.

Es wäre unter diesen Umständen schwierig gewesen, die Photosynthese mit chemischen Methoden zu untersuchen, wenn sich nicht gezeigt hätte, dass die Photosynthese aus zwei unabhängigenTeilreaktionen besteht, von denen nur die eine strukturempfindlich ist. Wie Ruben 1940 in Versuchen mit radioaktiver Kohlensäure fand, wird die Kohlensäure bei der Photosynthese zunächst in einer von selbst verlaufenden Dunkelreaktion gebunden, und diese Bindung der Kohlensäure—an eine hochmolekulare Substanz, deren Konstitution man noch nicht kennt—ist die strukturempfindliche Reaktion der Photosynthese.

Die andere Reaktion—die photochemische Reaktion, bei der die absorbierte Lichtenergie chemische Arbeit leistet—geht auch in den isolierten Chloroplasten, ja sogar in den isolierten Granula vor sich. Sie besteht in einer Zersetzung des Wassers, dessen Sauerstoff als gasförmiger Sauerstoff entwickelt wird. Der Wasserstoff des Wassers jedoch bleibt als starkes Reduktionsmittel gebunden und reduziert bei der Photosynthese die Kohlensäure der erwähnten hochmolekularen Kohlensäureverbindung.

Wasserstoffübertragende Fermente, von Otto Warburg, S. 366, Editio Cantor K.G., Aulendorf i. Württ.

LESSON 11

Constructions Involving Infinitives

A. With "zu"

B. Without "zu"

C. Modal Auxiliaries

A. THE INFINITIVE WITH "ZU"

1. Just as in English the infinitive is usually accompanied by the preposition *to* (to go, to show), so also in German an infinitive used as part of a sentence is generally accompanied by the preposition **zu**. The only exception to this rule so far has been the formation of the future tense and the introduction of an infinitive by a non-tense-forming (or "modal") auxiliary:

Wir werden (können, müssen) es **untersuchen.**
We will (can, must) *investigate* it.

as against

Wir versuchen (wagen, versprechen) es **zu untersuchen.**
We try (venture, promise) *to investigate* it.

The preposition **zu** always *immediately precedes* the verb itself. It even displaces detachable prefixes:

Wir versuchten, es noch einmal **zu** beschreiben.
We tried to describe it once more.

Wir versuchen, es auf einem anderen Weg durch**zu**führen.
We are trying to carry it out by another method.

2. The **zu**-infinitive, together with its object (if any) and its modifiers, makes up an *infinitive clause*. Since an infinitive is a constant verb form, and there is no variable verb form, the infinitive always comes at the very end of its clause. Long infinitive clauses are set off by commas. Shorter ones, especially those consisting of the **zu**-infinitive alone or of the **zu**-infinitive and merely a pronoun object, are not set off by commas. Other clauses are usually closed before the infinitive clause begins. The English translation may be an *-ing* form, or sometimes a subordinate clause introduced by the conjunction *that:*

Man ist imstande gewesen, eine Reihe von solchen Rechnungsregeln **abzuleiten**.
One has been able *to derive* a series of such arithmetical rules.

Die Aufgabe besteht darin **nachzuweisen**, daß . . .
The task consists in *demonstrating* that . . .

Man glaubte damals nicht an die Möglichkeit, das Atom **zu zerteilen**.
At that time they did not believe in the possibility *of dividing* the atom.

Er glaubte, darin etwas Wichtiges **zu erkennen**.
He thought (*that*) *he recognized* something important in that.

3. Three prepositions lend themselves to special use as conjunctions introducing infinitive clauses:

um . . . zu, in order to:

Um *x* und *y* zu finden, rechnen wir die linke Seite aus.
In order to find *x* and *y*, we calculate the left side.

anstatt . . . zu, instead of:

Anstatt Prüfungsmittel für seine Hypothese zu suchen, suchte er nur Beweismittel.

Instead of looking for material to test his hypothesis, he only looked for supporting evidence.

ohne . . . zu, without:

Ohne auf weitere Einzelheiten einzugehen, wollen wir hier das wesentliche zusammenfassen.

Without going into further details, we want to summarize the essentials here.

4. Infinitive introduced by the verb **sein** (to be):

Was ist (was war) daraus **zu schließen?**
What is (what was) *to be concluded* from that?

Note that the German **zu**-infinitive is translated by a *passive infinitive* ("to *be* concluded"). The combination "is to be concluded" may imply, according to context, "can be, must be *or* should be concluded." Generally speaking, it is best to leave the translation in the more nearly literal form unless the exact interpretation is absolutely clear or unless the tense of the verb "to be" is such that the more literal translation would be clumsy:

Was **wird** daraus zu schließen **sein?** (future)
What will we be able to (*or* will we have to) conclude from that? What will it be possible to conclude from that?

Was **wäre** daraus zu schließen? (short conditional)
What could (*or* might) be concluded from that?

B. THE INFINITIVE WITHOUT "ZU"

1. With "lassen"

There is another important instance where an infinitive, this time without *zu*, must be translated as a passive infinitive. This involves the verb **lassen** used as a reflexive verb and as an auxiliary:

Eine einfache Lösung **läßt sich** nicht ohne weiteres **finden.**
A simple solution *can* not *be* immediately *found*.
(*Literally:* . . . does not let itself be . . . found.)

Note that the following four wordings all mean practically the same thing:

Was kann man machen?	What can one do?
Was kann gemacht werden?	What can be done?
Was ist zu machen?	What is to be done?
Was läßt sich machen?	What can be done?

See also Section C, 2. below.

2. As an Imperative

The *infinitive without* "zu" is often used in giving operating instructions or directions for use:

Ventile vorsichtig aufmachen und nach Entleerung sofort schließen.

Open valves cautiously and close immediately after emptying.

C. MODAL AUXILIARIES

1. The "Full-Time" Modals

The modal auxiliaries are characterized by the fact that (*a*) they are used with the infinitive of another verb without the preposition zu, and (*b*) they have irregular forms:

INFINITIVE AND PAST PARTICIPLE	(BASIC MEANING)	PRESENT SINGULAR	PAST	SHORT CONDITIONAL
dürfen	(to be allowed to)	darf	durfte(n)	dürfte(n)
können	(to be able to)	kann	konnte(n)	könnte(n)
mögen	(to like to)	mag	mochte(n)	möchte(n)
müssen	(to have to)	muß	mußte(n)	müßte(n)
sollen	(to be supposed to)	soll	sollte(n)	sollte(n)
wollen	(to want to)	will	wollte(n)	wollte(n)

It is clear from this tabulation that:

(*a*). The infinitive and past participle of these verbs are iden-

Modal Auxiliaries

tical, hence all compound tenses are made up with the same constant form.

(*b*). The singular of the present tense involves an internal change (except for **soll**) and is devoid of the usual endings (ich, er, sie, es, man **darf**, etc.). (The plural of the present tense is regular, i.e., identical with the infinitive.)

(*c*). The past tenses all have a **-te** (or **-ten**) ending but no umlaut; the short conditional of the first four (alphabetically) has an umlaut, but **sollte(n)** and **wollte(n)** are never umlauted.

The perfect tenses of the modals are all formed with the auxiliary **haben**:

ich habe . . . dürfen, er hat . . . müssen, wir haben . . . können, etc.

Since the modals are used with the infinitive of another verb, any compound tense of the modals will bring together two forms both of which look like infinitives:

Ich habe es **sehen dürfen**. (**dürfen** is past participle)
I have been allowed to see it.

Er wird es **versuchen müssen**. (**müssen** is infinitive)
He will have to try it.

Such a *"double infinitive"* stands at the end of *any* clause, even, by way of exception, a dependent clause:

. . . , weil er es nicht rechtzeitig hat **versuchen können**.
. . . , because he has not been able to try it in time.

Note that the variable verb **hat** precedes the "double infinitive" in this dependent clause.

Meanings of the Modals. Many of the German modals have cognates, words of related linguistic origin, hence similar in form, in English:

> **müssen**, must, **mag**, may, **kann**, can
> **konnte**, could, **mochte**, might, **sollte**, should

But these English verbs are defective, i.e., they do not have a full set of tenses (*must* has no past, future, conditional, or perfects, *can* has a past but no future or perfects, etc.). Moreover, some of those that do have at least a present and a past sometimes involve a more profound difference in meaning between these two forms than merely one of time reference:

May—might, shall—should, will—would

This calls for considerable ingenuity on the part of the translator in adapting the basic meanings indicated above to the particular context.

(*a*). **Dürfen**

Man **darf** hieraus schließen, daß . . .
One *may* conclude from this that . . .
(Better:) It is permissible to conclude . . .

Man **darf** nicht schließen . . .
One *must not* conclude . . .

Das **dürfte** wahr sein
That *is probably* (or *might be*) true.

Dürfen denotes *having permission*, but the verb *may* is a dangerous translation because of its many other possible meanings. The *short conditional* sometimes implies probability.

(*b*). **Können**

Er **konnte** feststellen . . .   He *was able* to ascertain . . .
Er **könnte** feststellen . . .   He *would be able* to ascertain . . .
Er **kann** feststellen . . .   He *can* ascertain . . .

Können denotes ability, but the English verb *could* has two meanings, *was able* and *would be able*, and should therefore be used with caution in translating the past and conditional of the German verb.

(*c*). **Mögen**

Das **mag** sein. That *may* be.

Mag das günstig oder ungünstig sein . . .
May that be favorable or unfavorable . . .

Man **mochte** damals nicht experimentieren.
They *did* not *like* to experiment in those days.

Mögen has two distinct meanings: *may* and *like*. The former is by far the more frequent in scientific writing.

(d). Müssen

Die Wirkung **muß** (**mußte, müßte**) schneller erfolgen.
The action *must* (*had to, would have to*) take place faster.

Müssen indicates an obligation. Note that *only its present tense* can be translated *must*.

(e). Sollen

Die Arbeit **soll** morgen fertig sein.
The work *is supposed to* be (= *is expected to* be) finished tomorrow.

Die Arbeit **soll** dort sehr unangenehm sein.
The work *is supposed to* be (= *is said to* be) very unpleasant there.

In der vorliegenden Arbeit **soll** festgestellt werden, wie . . .
In der vorliegenden Arbeit **soll** man feststellen, wie . . .
In the present study it *is to* be ascertained (*or* one *is to* ascertain) how . . .

Was **sollen** wir tun?
What *shall* we (or *are* we *to*) do?

Es **sollten** noch viele weitere Elemente entdeckt werden. (*past*)
Many additional elements *were* still *to* be discovered.

Sollte diese Verbesserung zum Ziel führen, so **sollte** dieses Verfahren großen Erfolg haben (*short conditionals*).
Should this improvement lead to the goal, then this procedure *ought to* (or *should*) be a great success.

Sollen usually denotes a recognized obligation that may or may not be fulfilled. But its translation is often a very subtle matter.

(f). Wollen

Wir **wollen** dadurch nachweisen . . .
We *want* (i.e., *intend*) to demonstrate by this . . .

Es **will nicht** richtig funktionieren.
It *won't* (*will not*) work properly. (*not* a future!)

Wollen basically means *to want to* but also implies intention. Even though it can sometimes be translated *will,* it does *not* form future tenses!

CONDITIONAL PERFECT OF THE MODALS

Wir hätten es tun müssen.	We would have had to do it.
Wir hätten es tun wollen.	We would have wanted to do it.
Wir hätten es tun können.	We would have been able to do it.
or	We could have done it.
Wir hätten es tun sollen.	We should have done it.

Note especially the short English equivalents of the last two combinations above.

FULL-TIME MODALS USED ALONE. All of the modal auxiliaries listed above can occur *without* the infinitive of another verb, either because this infinitive is left understood or because the modal is used in special non-auxiliary meanings. In either case, all six modals form their perfects with a "regular" past participle:

gedurft, gekonnt, gemocht, gemußt, gesollt, gewollt.

Das Gas **muß** irgendwie hinaus.

The gas *has to get* out somehow. (A verb for "get" is left understood. This can be done whenever some adverb or separable prefix of *direction* or *motion* is included.)

Er **hat** nichts **gekonnt.**   He *could* not *do* anything.

Der Fremde **konnte** kein Deutsch (or **hat** kein Deutsch **gekonnt**).

The stranger *knew* no German. (Special use of *können* meaning *to know:* with languages only.)

2. Part-time Modals

A number of ordinary German verbs, some stem-changing and some not, lend themselves to use as modal auxiliaries, i.e., introduce the infinitive of another verb without **zu**. In that case these verbs adopt one of the chief characteristics of the full-time modals: instead of their usual past participle they use a form identical with their infinitive:

Ich habe es **gesehen.**   I saw it, I have seen it.
Ich habe es kommen **sehen.**   I saw it coming.

Other verbs that can be used as part-time modals are **hören** (to hear), **helfen** (to help), **heißen** (to bid, command), **lehren** (to teach), and especially **lassen, ließ, gelassen, läßt:**

Er hat einige Fehler durchgehen lassen.
He let a few mistakes slip through.
Er ließ seine Sekretärin mich anrufen.
He had his secretary call me.
Ich werde mir die Haare schneiden lassen.
I will have my hair cut.
Es hat sich noch keine einfachere Lösung finden lassen.
It has not yet been possible to find a simpler solution.

EXERCISE
BEMERKUNGEN ÜBER DIE NATUR DES MESOTRONS

Von H. Yukawa, Kyoto, Japan

Eingegangen am 9. Januar 1942

Einleitung. Bekanntlich hatte die Theorie des Mesotrons einen guten Erfolg betreffend die einheitliche Deutung der Kernkräfte, des β-Zerfalls und der Erscheinungen hinsichtlich der Höhenstrahlung im großen und ganzen; trotzdem stieß sie auf ernste Schwierigkeiten, als der eingehende quantitative Vergleich der Theorie mit der Erfahrung angestellt wurde. Es besteht kein Zweifel, daß es zweierlei Gründe für diese Schwierigkeiten gibt. Erstens haben wir die allgemeine Methode der relativistischen Quantentheorie, die von den wohlbekannten Divergenzschwierigkeiten ganz frei ist, noch nicht aufgefunden, und es ist sehr wahrscheinlich, daß die von der Erzeugung oder der Vernichtung der Mesotronen begleiteten Vorgänge völlig außerhalb des Anwendungsbereiches der gegenwärtigen Quantentheorie liegen, wie es schon von Heisenberg betont wurde. Zweitens aber ist es auch nicht sicher, ob die heutigen Voraussetzungen über die fundamentalen Eigenschaften, wie zum Beispiel den Spin und die Statistik der Elementarteilchen, alle richtig seien. Im Falle des Mesotrons ist dieser Punkt deshalb besonders zu beachten, weil der Spin und die Statistik desselben durch das Experiment allein nicht leicht bestimmt werden können. Daher ist es noch immer von großer Bedeutung, die Folgerungen der verschiedenen möglichen Voraussetzungen über die Natur des Mesotrons nacheinander zu prüfen, obwohl wegen des Mangels einer divergenzfreien Formulierung der Theorie die letzte Entscheidung heute unmöglich sein dürfte.

Das Problem des neutralen Mesotrons. Eine der wichtigsten Fragen, die wir noch nicht erwähnt haben, ist diejenige der Existenz neutraler Mesotronen. Ein von *Nishina* und *Birus* durchgeführter Versuch spricht gegen das Vorhandensein der neutralen Mesotronen in der Höhenstrahlung. Dies steht nicht notwendig in Widerspruch mit der Annahme, daß auch die neutralen Mesotronen verantwortlich für die Wechselwirkung zwischen Nukleonen seien, sondern das Nichtvorhandensein derselben in der Höhenstrahlung kann wohl auf ihre äußerst kurze Lebensdauer zurückgeführt werden, wie es von *Sakata* und *Tanikawa* theoretisch geschlossen wurde.

In dieser Beziehung ist zu bemerken, daß in der von *Marshak* entwickelten Mesotronenpaartheorie der Kernkräfte die neutralen Mesotronen nicht immer notwendig sind. Auch diese Theorie hat ebensoviele Vorteile wie Nachteile. Unter diesen ist besonders zu beachten, daß der Streuungsquerschnitt für das Mesotron mit Spin $1/2$ durchaus nicht klein, sondern sogar größer als derjenige für das gewöhnliche Mesotron ist. Weiter gibt es heute viele Versuchsergebnisse bezüglich des Mechanismus des Mesotronenzerfalls, die für die Annahme der Ganzzahligkeit des Mesotronenspins günstig sind. Jedoch können wir den folgenden Sachverhalt nicht übersehen.

> Excerpts from an article in *Zeitschrift für Physik,* Bd. 119, S. 201, Springer-Verlag, Berlin, 1942.

LESSON 12

Adjectives and Adverbs

As indicated in Lesson 4, Section B, it is possible to read and even to translate fairly accurately without knowing specifically what adjective endings are necessary in any particular situation. Yet the more serious student of German will find it helpful to master the principles of the adjective endings, which can sometimes be of help in interpreting parts of sentences, especially those involving extended modifier constructions:

Diese Erscheinungen sind offenbar psychologischer Natur.

These phenomena are evidently *of* a psychological nature.

. . . die zum Austritt sämtlicher Elektronen aus dem glühenden Draht notwendige Geschwindigkeit . . .

. . . the velocity necessary for the electrons to leave the incandescent wire . . .

The second example is merely a phrase taken out of a sentence. It helps to know that the ending on "notwendige" agrees with that on the initial article "die" before a feminine noun (recognized by the **-keit** suffix) and thus to locate the beginning and end of this unit.

A. THE ADJECTIVE ENDINGS

There are in actual fact only five adjective endings **-e**, **-em**, **-en**, **-er**, and **-es**. In order to interpret them it is necessary to look at each adjective-noun unit to see if it includes a *limiting adjective*. By definition, all **dieser**-words (including the definite article) and **ein**-words (see Lesson 4) are *limiting adjectives* and all other adjectives are classed as *descriptive*.

1. If the limiting adjective *has an ending*, concentrate on the limiting adjective to give you any necessary grammatical information regarding gender, number, and case. The descriptive adjective then always ends in **-e** or **-en** but these endings are of no grammatical significance:

MASCULINE	FEMININE	NEUTER
N. **der** neu**e** Prozeß	**die** neu**e** Methode	**das** neu**e** Verfahren
G. **des** neu**en** Prozesses	**der** neu**en** Methode	**des** neu**en** Verfahrens
D. **dem** neu**en** Prozesse	**der** neu**en** Methode	**dem** neu**en** Verfahren
A. **den** neu**en** Prozeß	**die** neu**e** Methode	**das** neu**e** Verfahren

PLURAL

N. **die** neu**en** Prozesse, Methoden, Verfahren
G. **der** neu**en** Prozesse, Methoden, Verfahren
D. **den** neu**en** Prozessen, Methoden, Verfahren
A. **die** neu**en** Prozesse, Methoden, Verfahren

2. If the limiting adjective has *no* ending, or if there is no limiting adjective at all, then the descriptive has information-giving endings (i.e., the endings a **dieser**-word would have if it were there).

ein neu**er** Prozeß	viele neu**e** Prozesse
a new process	many new processes
ein neu**es** Verfahren	mit kalt**em** Wasser
a new procedure	with cold water

In the genitive singular, masculine and neuter, the descriptive adjective always ends in **-en**:

des primär gebildeten Wassers
of the initially formed water

eine sekundäre Reaktion primär gebildeten Wassers mit Kohlenoxyd
a secondary reaction of initially formed water with carbon monoxide

B. THE EXTENDED MODIFIER CONSTRUCTION

Since every German attributive adjective must immediately precede the noun it modifies, any modifiers of the adjective must precede it, i.e., will be found between it and the limiting adjective, if any:

POSITION 1	POSITION 2	POSITION 3	POSITION 4
(limiting adjective)	(modifiers of the descriptive adjective)	(descriptive adjective)	(noun)
diese	**höchst**	**charakteristischen**	**Faktoren**
these	highly	characteristic	factors

In this setup, position 2 may also be filled by one or more prepositional phrases, but the presence of these will require a radical change in the English word order, a change essentially consisting in placing the descriptive adjective *after* the noun it modifies.

diese	**für die Strömung**	**charakteristischen**	**Faktoren**
these	for the flow	characteristic	factors

(these factors, characteristic of the flow)

der	**unter Umständen**	**starke**	**Strom**
the	under certain circumstances	strong	current

(the current, which is strong under certain circumstances)

It is this type of word grouping with modifiers preceding the descriptive adjective that is known as the *extended modifier construction*.

Since *participles* can be used as adjectives, they also enter into this construction; in fact they are more frequent in it than ordinary descriptive adjectives.

Past Participle

die	in diesem Kapitel	zusammengestellten	Grundsätze
the	in this chapter	assembled	principles

(the principles assembled in this chapter)

Present Participle (This is the form corresponding to the English *-ing* form of the verb. It is made up by adding a **-d** to the infinitive: **auftreten → auftretend**.)

eine	in solchen Fällen stets	auftretende	Erscheinung
a	in such cases always	occurring	phenomenon

(a phenomenon always occurring in such cases)

alle	dieser Bedingung*	genügenden	Werte
all	this condition	satisfying	values

(all values satisfying this condition)

die	die Funktionswerte	definierenden	Potenzreihen
the	the function values	defining	power series

(the power series defining the function values)

Future Participle (Formed by adding a **-d** to the **zu**-infinitive, whether the **zu** is inserted or a separate word)

die	vom Thermometer	abzulesenden	Temperaturen
the	from the thermometer	to be read off	temperatures

(the temperatures to be read off the thermometer)

die	in diesem Fall	zu erfüllenden	Bedingungen
the	in this case	to be fulfilled	conditions

(the conditions to be fulfilled in this case)

From the above examples it is evident that (*a*) the English is sometimes best rendered by means of a "which-clause" (*the*

---
* Dative case complement of the verb **genügen**.

current, which is strong, etc.); (*b*) present participles, even when used as adjectives, still retain enough of their verb nature to have objects and complements (*satisfying this condition, defining the function values*).

The chief difficulty presented by the extended modifier construction is that of recognizing it *as a unit* when it occurs in a sentence: recognizing its beginning and its end.

Its *beginning* is characterized by a limiting adjective immediately followed by either (*a*) a prepositional phrase or (*b*) another limiting adjective (provided there is no comma before the two successive limiting adjectives).*

Its *end* is characterized by a noun immediately preceded by either a descriptive adjective or a participle used as an adjective. This adjective can often be positively identified *by its ending*, i.e., is this ending the correct one to follow the one on the limiting adjective that starts the construction?

>   alle . . . genügend**en** Wert**e**
>   ein . . . stark**er** Strom

The translator must soon become accustomed to taking up the various parts or positions of these constructions in the order that will result in the best word order in English. One must proceed from position 4 *immediately* to position 3. (The best sequence for English translation is usually 1-4-3-2.)

The examples given above are rather simple ones chosen for purposes of illustration. In actual practice the extended modifier construction is often far more complex in that position 2 consists of *several* modifying words or phrases:

die/zum Austritt sämtlicher Elektronen aus dem glühenden Draht/notwendige/Geschwindigkeit.

the/for the exit of all electrons from the incandescent wire/ necessary/velocity.

(the velocity necessary for all the electrons to leave the incandescent wire)

---
* The comma would set off the words following it as a relative clause:
>   die Potenzreihen, **die die** Funktionswerte definieren
>   the power series, *which* define the function values

In cases like this it is often a matter of hunting for the noun that ends the construction after one has recognized a certain sequence (such as **die zum . . .** above) as its beginning. That **Geschwindigkeit** is the right noun is seen from the fact that it is in the right number (singular), and that it is not in a prepositional phrase and has no article or limiting adjective close before it. That **Austritt** cannot be the noun modified by the initial article **die** is clear from the fact that **Austritt** has its own article **dem** in the contraction **zum** (= **zu dem**); **Elektronen** is modified by an adjective that ends in -**er** and is thus *genitive* plural, while **Draht** has its own article **dem**. Such considerations may seem obvious, yet the beginning reader is all too prone to ignore a preposition like **zum** and blunder into combining **die** with **Austritt**, ignoring all details that should warn him to the contrary.

Extra Complications in the Extended Modifier Construction

1. Position 1 Vacant (i.e., no limiting adjective). This is particularly difficult to recognize in a full sentence:

Diese Darstellung gestattet es, **für die Praxis wichtige Daten,** wie Löslichkeit, Siedepunkt und Taupunkt, direkt abzulesen.

This representation makes it possible to read off directly *data important for practice,* such as solubility, boiling point, and dew point.

Sometimes this type of extended modifier construction can be recognized by a succession of two prepositions:

Auf Grund *von*   *in* neuerer Zeit   veröffentlichten   Ergebnissen
On the basis of   in recent times   published   results
(On the basis of results published in recent times)

2. Two or More Position-3 Adjectives Modifying the Same Noun. Each, or only one, of these adjectives may have position-2 modifiers. It must be noted that two or more adjectives modify-

ing the same noun are often separated by commas; *such a comma must not be mistaken for the end of a clause!*

ein neues, unabhängiges Verfahren
a new, independent technique

ein neues, von der Amplitudenmodulation unabhängiges Verfahren.
a new technique, independent of amplitude modulation

ein zu jener Zeit ganz *neues*, von der Amplitudenmodulation *unabhängiges* Verfahren
a technique quite new at that time and independent of amplitude modulation

Note that if an adjective that has an ending is followed by a comma, there must be another adjective with the same ending, and a noun for both adjectives to modify.

3. COMPLICATIONS IN POSITION 4. Modifying phrases following the noun in position 4. In translating, such phrases must be inserted before following up with position 3:

die dadurch   hervorgerufenen   Schwankungen **des Potentials**
the thereby   caused   fluctuations of the potential
(the fluctuations of the potential caused thereby)

das von *A* verbesserte Verfahren **von *B* zur unmittelbaren Chloridbestimmung** (2 phrases)
B's process for direct determination of chlorine, improved by *A*

4. ONE EXTENDED MODIFIER INSIDE THE OTHER.

Die in den den Kesseln zunächstliegenden Räumen befindlichen Leute . . .

The in the to the boilers lying closest rooms located employees . . .

The employees (located) in the rooms (lying) closest to the boilers . . .

In ihrer Arbeit nehmen diese Autoren an, daß der in der von ihnen zur Berechnung des von der elektrischen Feldeinheit induzierten Dipolmomentes früher angegebene Formel auftretende Frequenzmittelwert auch bei inhomogener Störung als gute Annäherung zu betrachten ist.

Note: If you can translate this unusually difficult and involved sentence correctly, you can probably stop reading this book. The correct English version is given in the Footnote.*

C. COMPARISON OF ADJECTIVES AND ADVERBS

1. Forms with Endings, i.e., Attributive Adjectives

The comparative degree (the "more-grade") of the adjective with endings is formed by adding **-er** to its stem; the superlative (the "most-grade") by adding **-st**. Both **-er** and **-st** are then followed by the usual grammatical endings **-e, -em, -en, -er,** or **-es:**

(a) **ein reiner Stein** a pure stone (positive)
  ein reinerer Stein a purer stone (comparative)
  mein reinster Stein my purest stone (superlative)
  ein reinster Stein a very pure stone (absolute superlative)

(b) **die reichliche Ausbeute** the abundant yield (positive)
  die reichlichere Ausbeute the more abundant yield (comparative)
  die reichlichste Ausbeute the most abundant yield (superlative)

Note that all German comparatives and superlatives are formed with *endings* and not with separate words like the English "more" and "most." The unwary reader may easily overlook the presence of these endings. Their recognition, especially that of the comparative **-er**, is rendered all the more difficult by the fact that some adjectives have stems ending in **-er**.

sicher (sure) **ein sicherer Beweis** a sure proof
  ein sichererer Beweis a surer proof

---

* In their study these authors assume that the mean value of the frequency appearing in the formula specified by them for the calculation of the dipole moment induced by the field unit is to be regarded as a good approximation even with inhomogenous interference.

teuer (expensive) **teure*** **Vorrichtungen** expensive devices
　　　　　　　　　teurere Vorrichtungen more expensive devices

2. *Forms without Endings, i.e., Adverbs and Predicate Adjectives*

The comparative degree is formed by adding **-er**. The superlative degree is *a phrase* beginning with the prepositional contraction **am** and ending in **-sten**:

ADVERB.

A läuft **schnell**.	A runs fast	(positive)
B läuft **schneller**.	B runs faster	(comparative)
C läuft **am schnellsten**.	C runs fastest	(superlative)

PREDICATE ADJECTIVE.

Methode A ist leicht.	Method A is easy	(positive)
Methode B ist leichter.	Method B is easier	(comparative)
Methode C ist am leichtesten.	Method C is easiest	(superlative)

die **am leichtesten** in Alkohol löslichen Stoffe.
the most easily　　　 in alcohol　 soluble　　 substances.
(the substances most easily soluble in alcohol).

3. *Irregular Comparatives and Superlatives*

Most of the more common adjectives form their comparative and superlative with umlaut:

　　lang,　　**länger,**　　**am längsten,**　　**der längste**
　　kurz,　　**kürzer,**　　**am kürzesten,**　　**der kürzeste**

Some are irregular:

(large)	**groß**	**größer**	**der größte**	
(good, well)	**gut**	**besser** (better)	**der beste**	(best)
(high)	**hoch**	**höher**	**der höchste**	
(near)	**nah**	**näher**	**der nächste**	(nearest, next)
(much)	**viel**	**mehr** (more)	**das meiste**	(most)

* An **e** before an **r**, if preceded by a vowel, is dropped before an ending beginning with **-e**!

Note. die meisten Leute — most people
mehr Geld — more money

Do not confuse **mehr** (more) with **mehrere** (-r, -n) (several).
But **ein oder mehrere** is equivalent to "one or more."

4. Special Uses of the Comparative

(a) eine längere Diskussion — a *fairly long* discussion
immer länger — longer and longer
immer bedeutender — more and more significant
immer weiter — further and further, on and on
immer wieder — again and again

(b) PARALLEL COMPARATIVES
je größer, je besser
the bigger the better

Je größer die Drucksenkung ist, desto mehr wird die Entropie zunehmen.

The greater the pressure drop, the more the entropy will increase.

Bei gleicher Temperatur ist die Emission um so* größer, je geringer die Austrittsarbeit ist.

At equal temperature the emission is so much the greater, the smaller the work function (is).

5. Endingless Superlatives

Note the use of the form **möglichst** without an ending:
möglichst vollständig
as complete as possible
die möglichst vollständige Aufarbeitung
the most complete recovery possible

A number of other "endingless" superlatives (**meist, höchst, äußerst**) have special meanings. Consult your dictionary.

---

* Often spelled **umso**, though never listed this way in dictionaries.

6. Comparison of Equals

A ist so groß **wie** B
A is *as* great *as* B

But note:

A **sowie** B   A as well as B

and:

sowohl die großen $\begin{Bmatrix} \text{wie} \\ \text{als} \end{Bmatrix}$ (auch) die kleinen

both the large ones and the small ones
the large ones as well as the small ones

D. ADJECTIVES USED AS NOUNS

die Reichen und die Armen	the rich and the poor (*plur.*)
das Alte und das Neue	the old and the new (*sing.*)
der Arme, ein Armer	the poor man, a poor man
die Arme	the poor woman
der Beschuldigte	the accused (man)
der Versicherte, ein Versicherter	the insured, an insured man
die Versicherte	the insured (woman)
die Versicherten	the insured (persons)
Versicherte	insured people
Altes und Neues	old things and new things
etwas Neues	something new

Note that in German an adjective used as a noun (*a*) is capitalized; (*b*) retains its adjective endings and uses them to specify gender and number in a way that can be rendered in English only by using nouns like *man, woman, person, people* or *things*.

Some concepts represented by nouns in English are represented in German by adjectives used as nouns:

der Fremde, die Fremde	the stranger (*m.* and *f. respectively*)
ein Bekannter, eine Bekannte	an acquaintance
die Angestellten	the employees

In others the gender is fixed by the unexpressed noun that originally followed:

die Konstante	(= die konstante Zahl)	the constant
eine Gerade	(= eine gerade Linie)	a straight line

Great flexibility and economy of expression is often achieved by the use of participles as nouns:

der Genannte the aforementioned (person)

Ein Geübter kann das sehr schnell fertigbringen.
A practiced expert can accomplish this very quickly.

die Fliehenden the fleeing people, the fugitives

Er erzählte **das Geschehene**.
He related *what had happened*.

das in Abschnitt II **Angenommene**
what was assumed in section II

das für diesen Beweis **Gegebene**
the *"given"* (= the given data) for this proof

If the noun understood after the adjective has already been mentioned and is omitted merely to avoid repetition, the adjective is not capitalized. This is often equivalent to using the word *one* in place of the noun in English:

Unter den physikalischen Theorien gibt es leichtere und schwierigere.

Among the physical theories there are easier ones and more difficult ones.

Sometimes the "understood" noun is one yet to be mentioned:

Es werden einerseits vorwiegend leichtsiedende flüssige oder auch gasförmige, auf der anderen Seite vor allem hochsiedende feste Kohlenwasserstoffe erzeugt.

On the one hand predominantly low-boiling liquid or even gaseous hydrocarbons are produced, on the other hand above all high-boiling solid ones.

All of the adjectives ending in **-e** modify the one noun *hydrocarbons;* the **-e** ending serves the reader as a warning that there must be a noun either present or understood. In English it is best to put in the noun after the first set of adjectives and use the pronoun "ones" to replace it after the second set.

Adverbial Phrases with IM and AM

im allgemeinen	in general, generally
im wesentlichen	essentially
im einzelnen	in detail, in particular
im ganzen	on the whole

The student should be able to recognize these phrases so that he will not try to look for a noun after the forms with the -en ending. Remember also that phrases like **am leichtesten** are superlatives of "endingless" adjective forms (See Section C, 2 of this lesson).

EXERCISE
ZUR ELEKTRODYNAMIK BEWEGTER KÖRPER
Von A. Einstein

Wollen wir die *Bewegung* eines materiellen Punktes beschreiben, so geben wir die Werte seiner Koordinaten in Funktion der Zeit. Es ist nun wohl im Auge zu behalten, daß eine derartige mathematische Beschreibung erst dann einen physikalischen Sinn hat, wenn man sich vorher darüber klar geworden ist, was hier unter „Zeit" verstanden wird. Wir haben zu berücksichtigen, daß alle unsere Urteile, in welchen die Zeit eine Rolle spielt, immer Urteile über *gleichzeitige Ereignisse* sind. Wenn ich z.B. sage: „Jener Zug kommt hier um 7 Uhr an," so heißt dies etwa: „Das Zeigen des kleinen Zeigers meiner Uhr auf 7 und das Ankommen des Zuges sind gleichzeitige Ereignisse." [1]

Es könnte scheinen, daß alle die Definition der „Zeit" betreffenden Schwierigkeiten dadurch überwunden werden könnten, daß ich an Stelle der „Zeit" die „Stellung des kleinen Zeigers meiner Uhr" setze. Eine solche Definition genügt in der Tat, wenn es sich darum handelt, eine Zeit zu definieren ausschließlich für den Ort, an welchem sich die Uhr eben befindet; die Definition genügt aber nicht mehr, sobald es sich darum

---

[1] Die Ungenauigkeit, welche in dem Begriffe der Gleichzeitigkeit zweier Ereignisse an (annähernd) demselben Orte steckt und gleichfalls durch eine Abstraktion überbrückt werden muß, soll hier nicht erörtert werden.

handelt, an verschiedenen Orten stattfindende Ereignisreihen miteinander zeitlich zu verknüpfen, oder—was auf dasselbe hinausläuft—Ereignisse zeitlich zu werten, welche in von der Uhr entfernten Orten stattfinden.

Wir könnten uns allerdings damit begnügen, die Ereignisse dadurch zeitlich zu werten, dass ein samt der Uhr im Koordinatenursprung befindlicher Beobachter jedem von einem zu wertenden Ereignis Zeugnis gebenden, durch den leeren Raum zu ihm gelangenden Lichtzeichen die entsprechende Uhrzeigerstellung zuordnet. Eine solche Zuordnung bringt aber den Übelstand mit sich, daß sie vom Standpunkte des mit der Uhr versehenen Beobachters nicht unabhängig ist, wie wir durch die Erfahrung wissen. Zu einer weit praktischeren Festsetzung gelangen wir durch folgende Betrachtung.

Befindet sich im Punkte A des Raumes eine Uhr, so kann ein in A befindlicher Beobachter die Ereignisse in der unmittelbaren Umgebung von A zeitlich werten durch Aufsuchen der mit diesen Ereignissen gleichzeitigen Uhrzeigerstellungen. Befindet sich auch im Punkte B des Raumes eine Uhr—wir wollen hinzufügen, „eine Uhr von genau derselben Beschaffenheit wie die in A befindliche"—so ist auch eine zeitliche Wertung der Ereignisse in der unmittelbaren Umgebung von B durch einen in B befindlichen Beobachter möglich. Es ist aber ohne weitere Festsetzung nicht möglich, ein Ereignis in A mit einem Ereignis in B zeitlich zu vergleichen; wir haben bisher nur eine „A-Zeit" und eine „B-Zeit", aber keine für A und B gemeinsame „Zeit" definiert. Die letztere Zeit kann nun definiert werden, indem man *durch Definition* festsetzt, dass die „Zeit", welche das Licht braucht, um von A nach B zu gelangen, gleich ist der „Zeit", welche es braucht, um von B nach A zu gelangen.

(Fortsetzung folgt)

LESSON 13

Prepositions

The reader of German soon learns to recognize such *phrases* as the following by their structural similarity to English phrases:

in diesen Untersuchungen	in these investigations
für solche Versuche	for such experiments
mit starken Detektoren	with strong detectors
durch Entartung	by (means of) degeneration
bei den Photonen	in the case of photons

The words introducing these phrases and linking them to the rest of their sentences are known as *prepositions*. The language student will soon realize that these words have little or no definite meaning outside of the particular phrases in which they occur in any given context.* In this lesson certain general or usual meanings are indicated for some prepositions, but these can not be expected to cover all possible contexts.

In German, prepositions present two problems:

1. They must be recognized and kept apart from other sen-

---

* This is as true in English as in German. Try, for instance, to define the meaning of the word *to*, or observe the use of *with* in these sentences:

Are you *with* me or against me?	(with *vs.* against)
Don't fight *with* this enemy.	(with = against)

tence elements that resemble them, especially separable prefixes. Remember: a preposition always introduces a phrase; therefore it cannot ordinarily come at the end of a clause. Conversely, an apparent preposition at the end of a clause should usually be construed as a separable prefix.

Das Wasser fließt **durch** den Schlauch in die Wanne. (*preposition*)

The water flows *through* the tube into the trough.

Dann fließt Wasser **durch.**

Then water flows *through.* (*prefix:* verb **durchfließen**)

2. They "take" cases, i.e., their objects in the phrases must be in the genitive, dative, or accusative case according to the very arbitrary principle that certain prepositions require objects in certain cases. While the exact case is often of little importance to the reader, it is sometimes of value in identifying phrases as units, especially in threading one's way through long and involved sentences. For this reason a table listing all important prepositions with approximate meanings, grouped according to the cases they "govern," is given in Appendix 2.

"POSTPOSITIONS"

Some German prepositions are in actual practice "postpositions," i.e., they often, or always, follow their object instead of preceding it, or to put it in different terms, they end their phrases instead of beginning them:

Wegen dieses Umstandes } because of this circumstance
dieses Umstandes wegen

nach dieser Theorie } according to* this theory
dieser Theorie nach

Der Einfachheit halber
for the sake of simplicity

(Seine Haltung) dieser Meinung gegenüber
(His attitude) toward this opinion

---

* **Nach** as a "postposition" occurs only in this meaning.

Der Erfahrung gemäß
In accordance with experience

Some form combinations with their objects, especially with pronoun objects:

krankheitshalber	on account of illness
demgegenüber	compared to that
demnach, demgemäß	according to that, accordingly
deswegen	for that reason, on that account
dessenungeachtet	notwithstanding
trotzdem	in spite of that, nevertheless
außerdem	besides, in addition to that

Some of these combinations are very simliar to the **da**-words, **hier**-words and **wo**-words presented in Lesson 5 C and D.

"DOUBLE PREPOSITIONS"

Some prepositions introduce phrases with an adverb (or what often appears to be another preposition) at the end:

von dieser Zeit an (or **ab**)   from this time on.

But often the translation is only one word, a preposition:

über dieses Maß hinaus	beyond this measure
von diesem Gesichtspunkt aus	from this point of view
über längere Zeitperioden hin	over longer periods of time
von der äußeren Zone her	from the outer zone

Die Winkelgeschwindigkeit nimmt vom Mittelpunkt **nach außen zu** ab.
The angular velocity decreases from the center *outward*.

The preposition **bis** is, more often than not, followed by **an, auf, in,** or **zu,** which then determines the case of the object.

bis an diesen (or **bis zu diesem**) **Punkt**   up to this point.

"**bis auf**" has two meanings:

alle bis auf diesen	all except (for) this one
Es reicht bis auf diesen.	It extends up to this one.

Contractions

It is regular practice, even in very formal writing, to contract prepositions into a single word with the definite article forms **das**, **dem** and (in one instance only) **der**:

ins Wasser	= in das Wasser	into the water
ans Ende	= an das Ende	to the end
durchs Eis	= durch das Eis	through the ice
im Wasser	= in dem Wasser	in the water
am Ende	= an dem Ende	at the end
beim Erhitzen	= bei dem Erhitzen	on heating
vom Felde	= von dem Felde	from the field
zum Erhitzen	= zu dem Erhitzen	for heating
vorm Erhitzen	= vor dem Erhitzen	before heating
zur Ausgleichung	= zu der Ausgleichung	for the equalization

Prepositions versus Separable Prefixes

One of the reader's greatest sources of error and annoyance is his natural tendency to confuse prepositions and separable prefixes. Even though "postpositions" and "double prepositions" tend to add to the difficulty, the best guiding principle is still the one intimated at the beginning of this lesson: *a preposition must be in a prepositional phrase,* and in the vast majority of cases it will begin its phrase. A word that looks like a preposition occurring at the end of a clause should always be construed as a separable prefix unless the context makes this impossible and forces one to look for another possibility ("postposition" or second part of a "double preposition"):

Nur wenige betrachteten es von diesem Gesichtspunkt **aus**.
Only few people viewed it from this point of view.
(**von . . . aus** "double preposition")

Das schließt die Möglichkeit solcher Fehler **aus**.
This excludes the possibility of such errors.
(**ausschließen** separable verb)

Was für ein Schluß geht **aus** diesen Tatsachen hervor?
What sort of conclusion is evident from these facts?
(**aus diesen Tatsachen** prepositional phrase)

Many verbs in both English and German are habitually used with certain prepositions, that is, with *phrases* introduced by certain prepositions. Thus the verb *think* is often followed by a phrase beginning with *of:*

Think *of* the possibilities.
Man denke **an** die Möglichkeiten.

Teachers and students alike will often refer to this by saying "**denken an** means *to think of.*" But remember that **an** is a preposition introducing a phrase and not by any means a separable prefix. In other words, don't expect to find it listed as *andenken* in the dictionary!

The verb involved in such a combination may actually be a separable verb, but will then have a prefix of its own:

Es **hängt** von der Temperatur **ab**.
Es **kommt** auf die Temperatur **an**.
It depends on the pressure.

The verbs involved here are **abhängen** and **ankommen**.

The first is used with a phrase beginning with **von** to mean *on*, the second with **auf** at the beginning of the phrase.

In dealing with prepositions it is a cardinal rule *never* to break up a prepositional phrase, that is, always to treat it as a unit and never try to translate its object without also translating the preposition. This is particularly important in dealing with extended modifier constructions:

die	von Photokathoden	bei der Belichtung	ausgelösten
the	by photo-cathodes	on exposure to light	released
Elektronen			
electrons			

In identifying the noun **Elektronen** as the end of the construction and as the one modified by the initial **die**, it helps to realize that the other two nouns are in prepositional phrases and must not be taken out of them, hence cannot be construed with **die**.

Only in one situation can a prepositional phrase be broken up in translation, and that is the case of subjectless passive, as described in Lesson 9, Section B, 2.

EXERCISE
ZUR ELEKTRODYNAMIK BEWEGTER KÖRPER

Von A. Einstein

(Fortsetzung)

Es gehe nämlich[1] ein Lichtstrahl zur „A-Zeit" t_A von A nach B ab, werde zur „B-Zeit" t_B in B gegen A zu reflektiert und gelange zur „A-Zeit" t'_A nach A zurück. Die beiden Uhren laufen definitionsgemäß synchron, wenn

$$t_B - t_A = t'_A - t_B.$$

Wir nehmen an, daß diese Definition des Synchronismus in widerspruchsfreier Weise möglich sei, und zwar[2] für beliebig viele Punkte, daß also allgemein die Beziehungen gelten:

1. Wenn die Uhr in B synchron mit der Uhr in A läuft, so läuft die Uhr in A synchron mit der Uhr in B.

2. Wenn die Uhr in A sowohl mit der Uhr in B als auch mit der Uhr in C synchron läuft, so laufen auch die Uhren in B und C synchron relativ zueinander.

Wir haben so unter Zuhilfenahme gewisser (gedachter) physikalischer Erfahrungen festgelegt, was unter synchron laufenden, an verschiedenen Orten befindlichen, ruhenden Uhren zu verstehen ist und damit offenbar eine Definition von „gleichzeitig" und „Zeit" gewonnen. Die „Zeit" eines Ereignisses ist die mit dem Ereignis gleichzeitige Angabe einer am Orte des Ereignisses befindlichen, ruhenden Uhr, welche mit einer bestimmten, ruhenden Uhr, und zwar für alle Zeitbestimmungen mit der nämlichen[3] Uhr, synchron läuft. Wir setzen noch der Erfahrung gemäß fest, daß die Größe

$$\frac{2\overline{AB}}{t'_A - t_A} = V$$

eine universelle Konstante (die Lichtgeschwindigkeit im leeren Raume) sei.

Wesentlich ist, daß wir die Zeit mittels im ruhenden System ruhender Uhren definiert haben; wir nennen die eben definierte Zeit wegen dieser Zugehörigkeit zum ruhenden System „die Zeit des ruhenden Systems".

§ 2. Über die Relativität von Längen und Zeiten

Die folgenden Überlegungen stützen sich auf das Relativitätsprinzip und auf das Prinzip der Konstanz der Lichtgeschwindigkeit, welche beiden Prinzipien wir folgendermaßen definieren.

1. Die Gesetze, nach denen sich die Zustände der physikalischen Systeme ändern, sind unabhängig davon, auf welches von zwei relativ zueinander in gleichförmiger Translationsbewegung befindlichen Koordinatensystemen diese Zustandsänderungen bezogen werden.

2. Jeder Lichtstrahl bewegt sich im „ruhenden" Koordinatensystem mit der bestimmten Geschwindigkeit V, unabhängig davon, ob dieser Lichtstrahl von einem ruhenden oder bewegten Körper emittiert ist. Hierbei ist

$$\text{Geschwindigkeit} = \frac{\text{Lichtweg}}{\text{Zeitdauer}},$$

wobei „Zeitdauer" im Sinne der Definition des § 1 aufzufassen ist.

Excerpts from *Annalen der Physik*, IV. Folge, 17, pp. 891–895, Johann Ambrosius Barth, Leipzig, 1905.

Notes for Exercise

[1] Start your sentence with this word "nämlich", translating it "that is."
[2] **und zwar** and as a matter of fact, and to be specific.
[3] As an adjective, the word means "same"!

APPENDIX **1**

Details Regarding Demonstratives

In German the demonstratives appear in a great variety of forms and are subject to all the gender, number, and case variations of the noun-adjective system (Lesson 4).

In scientific writing those of the **der-die-das** type appear chiefly as the antecedents of phrases (especially possessive phrases) and relative clauses:

der Radius der Röhre und **der** des Heizfadens.
the radius of the tube and *that* of the hot filament.

die Ladung des Kerns und **die** in der äußersten Schale.
the charge of the nucleus and *that* (or *the one*) in the outermost shell.

der gefundene Punkt und **der,** der noch zu finden ist.
the point found and *the one* that is still to be found.

gerade in **d e n** Fällen, die wir untersuchen.
precisely in *those* cases that we are investigating.

Das ist ein Hüttenbauverfahren, das **dem** der Stahlindustrie gleicht.
That is a metallurgical process that resembles *that* of the steel industry.

Die Schwierigkeit ist **die**, daß man es mit zuvielen Variablen zu tun hat.

The difficulty is *this*, that we are dealing with too many variables.

The genitive case forms **dessen** and **deren** are used to mean "the latter's," to avoid ambiguity with possessive adjectives:

Theoretisch müßten diese Elektronen zur Anode fliegen, sobald **deren** Potential den Wert Null überschritten hat.

Theoretically these electrons would have to fly toward the anode, as soon as its (*or* the latter's) potential has exceeded the value zero.

"*Its* potential" is clear enough in English, but in German **ihr Potential** would be ambiguous, since **ihr** could mean either "its" or "their," the latter referring to the plural "**die Elektronen.**"

The **dieser**-words **dieser** (this) and **jener** (that) are the most obvious demonstratives, yet they sometimes present difficulties in translation, especially when used as pronouns. **Dieser** is by far the more frequent, and often means simply *it* or, if plural, *they* or *them*, besides the more literal *this* or *these:*

Prallen nun Elektronen auf die Luftmoleküle auf, so werden **diese** in positive Ionen und negative Elektronen zerspalten.

Now, if electrons strike against the air molecules, then *these* (or *the latter*) are split up into positive ions and negative electrons.

Sie erhielten erst Kenntnis von der Gefahr, in welcher sie geschwebt hatten, als **diese** bereits beseitigt war.

They did not receive knowledge of the danger they had been in until *it* had already been eliminated. (**diese** refers to **die Gefahr**, the danger)

The forms **dies** and **das** are often used as subject or object pronouns. As subjects they can be followed by plural verb forms:

Dies gilt besonders für die Metalle.
This is especially true of metals.

Das sind Faktoren, die bisher vernachlässigt wurden.
Those are factors that were neglected up to now.

Dieser and **jener** often mean "the latter" and "the former."

Wir haben hier eine negative und eine positive Elektrode. Jene wird die Kathode und diese die Anode genannt.

We have here a negative and a positive electrode. The former is called the cathode and the latter the anode.

The words **ersterer** and **letzterer** are used in similar meanings. The second sentence of the above example could also be worded:

Erstere wird die Kathode und letztere die Anode genannt.

The word **solcher** is also used as a demonstrative:

Wir haben zweierlei Vektoren kennengelernt, die als Funktionen des Ortes Vektorfelder bestimmen, **solche**, deren Divergenz Null, und **solche** deren Rotation Null ist.

We have become acquainted with two different kinds of vectors determining vector fields as functions of place, *those* whose divergence is zero and *those* whose rotation is zero.

Der Sauerstoff hat ein spez. Gewicht von 1,1052,* der Stickstoff **ein solches** von 0,9713.*

Oxygen has a specific gravity of 1.1052,* nitrogen has *one* of 0.9713.*

The word **welcher** also has special uses as a pronoun; it is the opposite of **keine(s)** (none):

Es gibt **welche**, die viel schneller verlaufen.

There are *some* that proceed much faster.

Two German demonstratives have the peculiarity of being combinations of the definite article and an adjective, somewhat like the English solid spelling of *another* (= an other):

derselbe,	**dieselbe,**	**dasselbe**	the same
derjenige,	**diejenige,**	**dasjenige**	that, the (very) one

The first component of these undergoes all the variations of the definite article, while the second component then has the requisite adjective endings **-e** or **-en**:

---

* See Appendix 4.

Details Regarding Demonstratives

dieselben (pl.), **demselben, demjenigen,** etc.

The word **derselbe** is often used as a pronoun, somewhat in the manner of the official-legal use of *same* in English in such messages as:

Sighted sub, sank same,

where *same* simply means *it*. The simple pronouns *it, they, them* will thus usually be the best translation of the German:

Führt man dem Motor eine höhere Spannung bei gleicher Frequenz zu, so erhöht sich die Drehzahl **desselben** nicht.

If a higher voltage is fed to the motor at the same frequency, the rpm *of it* (*or its* rpm) is not increased.

The word **derjenige** exists solely for the purpose of anticipating phrases and relative clauses:

diejenigen Metalle, die eine Jahresproduktion von mehreren hunderttausend Tonnen aufweisen

those metals that show an annual production of several hundred thousand tons

Die Menge an A scheint **diejenige** an B etwas zu übertreffen.
The quantity of A seems to exceed *that* of B somewhat.

APPENDIX 2

Prepositions and Their Cases

1. Governing the Genitive Case

This list is not complete because of the large number of rather infrequent prepositions that would otherwise have to be included. Many are based on nouns or adjectives, such as the last three in the list.

anstatt *or* statt	instead of
trotz	in spite of
während	during
wegen	because of, on account of
um . . . willen*	for the sake of
außerhalb	outside (of)
innerhalb	inside (of)
oberhalb	above
unterhalb	below, underneath
diesseits	this side of
jenseits	that side of, beyond
halber	for the sake of; because of
kraft	by virtue of
längs	along

* The object comes between the two parts of the preposition: **um des Reimes willen** for the sake of rhyme.

laut	by virtue of, according to
mittels(t)	by means of
einschließlich	including, inclusive of
ausschließlich	excluding, exclusive of
angesichts	in view of

2. Governing the Dative Case Only

aus	out of, from, (made) of
außer	outside of, except for, in addition to
bei	in the case of, with, in, in view of
dank	thanks to, owing to
gegenüber	opposite, toward, in comparison with
gemäß	according to, in accordance with
mit	with, by
nach	after; to, toward; according to
seit	(ever) since, for (*a period of time*)
von	of, from; by (*with passives*)
zu	to, toward; for (the purpose of)

3. Governing the Accusative Case Only

bis	until, till, up to
durch	through, by (means of), as a result of
für	for
gegen	against, toward, compared to
ohne	without
um	about, around; by (*a quantity*)
wider	against

4. Governing Both Dative and Accusative

an	to, at, by
auf	on, to, for
hinter	behind, in back of
in	in, into
neben	beside, alongside of, in addition to
über	over, above, across; (*accusative only*) about, concerning

unter	under, below; among
vor	before, in front of; for (*fear, joy, etc.*); (*dat. only*) ago*
zwischen	between

The prepositions in Group 4 govern the accusative case when motion or aim at a specific goal is implied, and the dative when location or position are implied.

in das Wasser (*accusative*)   into the water
in dem Wasser (*dative*)   in the water

Liegt x über diesem Wert . . . (*dative*)
If x lies above this value . . .

Steigt x über diesen Wert (hinaus) . . . (*accusative*)
If x rises above (*or* beyond) this value . . .

---

* **vor einiger Zeit** some time ago.

APPENDIX **3**

The Alphabet

German is now printed almost entirely in Roman type, but since many books printed in Gothic are still current, the Gothic type is shown here also, even though it has not been used in scientific articles for many decades.

The forms with the **e** are makeshifts used whenever the umlaut mark is not available in Roman type or on the typewriter. They should be avoided if at all possible.

In recognizing words of non-German origin, which are often derived from the same Latin or Greek sources as the corresponding English words, it is helpful to know that:

(*a*). **c** is usually replaced by **k** or **z** (because it is pronounced like one or the other of these):

konzentrieren, instead of **concentrieren,** to concentrate

(*b*). **ä** and **ö** are often equivalent to English **ae, oe,** or **e**:

Äther ether   **ästhetisch** aesthetic
ökonomisch economic **Homöopathie** homoeopathy
                                         homeopathy

The Alphabet

CAPITALS		SMALL LETTERS		CAPITALS		SMALL LETTERS	
ROMAN	GOTHIC	ROMAN	GOTHIC	ROMAN	GOTHIC	ROMAN	GOTHIC
A	𝔄	a	a	O	𝔒	o	o
B	𝔅	b	b	P	𝔓	p	p
C	ℭ	c	c	Q	𝔔	q	q
		ch	ch	R	𝔑	r	r
		ck	ck	S	𝔖	s	s
D	𝔇	d	d				s[2]
E	𝔈	e	e			ß	ß
F	𝔉	f	f			st	st
G	𝔊	g	g	T	𝔗	t	t
H	𝔍	h	h			tz	tz
I	𝔍	i	i	U	𝔘	u	u
J	𝔍	j	j	V	𝔙	v	v
K	𝔎	k	k	W	𝔚	w	w
L	𝔏	l	l	X	𝔛	x	x
M	𝔐	m	m	Y	𝔜	y	y
N	𝔑	n	n	Z	𝔝	z	z

UMLAUT

Ä, Ae 𝔄e ä, ae ä, ae   Ü, Ue 𝔘e ü, ue ü, ue
Ö, Oe 𝔒e ö, oe ö, oe   Äu, Aeu 𝔄u äu, aeu äu, aeu

[1] At the beginning or in the middle of a syllable only.
[2] At the end of a syllable.

APPENDIX **4**

Numbers

Since numbers are usually set in figures in German and rarely appear written out in words, they present little difficulty in reading, except perhaps the following details:

A. Numbers, when written out, often appear as one word:

hundert/neun/und/zwanzig
(hundred nine and twenty)   one hundred twenty-nine

B. Ordinal numbers are:

1. Adjectives formed by adding **-t** or **-st** (plus an adjective ending) to the cardinals:

         neunzehn   **der neunzehnte Dezember**
         nineteen   the nineteenth of December

2. Abbreviated by means of a *period:*

         am 22. Mai = am zweiundzwanzigsten Mai
                    on the twenty-second of May

C. Punctuation:

1. A *comma* stands for a decimal point:

GERMAN	ENGLISH
3,1416	3.1416
0,002	0.002

(The German numbers are read "drei Komma eins vier eins sechs" and "null Komma null null zwei.")

2. *Spacing* is used to set off thousands:

GERMAN	ENGLISH
3 796 000	3,796,000

APPENDIX **5**

Most Frequent Words

The following is an alphabetical list of approximately 200 words most frequently encountered in reading German prose,* together with their most common English equivalents. They account for practically half or more of the vocabulary of any ordinary reading passage. It must be kept in mind, however, that many of the words occurring most often are the very ones that either have multiple uses or meanings or, if they are prepositions, are particularly hard to restrict to any specific meaning and translation. Asterisks indicate stem-changing verbs; a superscript ˢ (as in ausˢ) indicates that the word also occurs as a separable prefix. Verbs are listed in the infinitive form only; adjective entries include adverbial uses.

ab^s, off, away
aber, but, however
all, all
als, as, when, than

also, therefore, so
alt, old
an^s, on, at, to, etc.
ändern, to change

---

* Based on B. Q. Morgan's *German Frequency Word Book,* 1928, prepared by the Committee on Foreign Languages, American Council on Education, with a few omissions to adapt the list to the "scientific" reading vocabulary.

arbeiten, to work
die Art, type, manner, way
auch, also, even
auf[s], on, up, to, etc.
das Auge, eye
aus[s], out (of), from

bei[s], in, at, with, in the case of
beide, both, two
besser, better
bis, until, up to
bleiben,* to stay, remain
blicken, to look, glance
binden,* to bind, tie
bringen,* to bring, put, take

da, 1. *adv.* there; then
    2. *conj.* since
dann, then
daß, that (*conj.*)
denken,* to think
denn, for (*conj.*)
der, die, das, the
deutsch, German
dienen, to serve, be used
dieser, this, *plur.* these
doch, yet, still, after all
drei, three
durch[s], through, by means of

eben, *adv.* just; simply
eigen, own
ein[s], *sep. pref.* in, into
ein(s), a, an, one
einzeln, individual
enden, to end
er, he, it
erst, first, only, not until
es, it

fahren,* to go, ride, drive
fallen,* to fall
fassen, to grasp, express
fern, far, distant
finden,* to find
folgen, to follow, obey
fort[s], on, away
fragen, to ask
frei, free
der Freund, friend
führen, to lead, carry
für, for

ganz, whole, entire, quite
geben,* to give
gegen, against, toward
gehen,* to go
der Geist, spirit
gemein, common
das Gesetz, law
glauben, to believe
gleich, equal, same; *adv.* right, immediately
das Glück, luck, happiness
der Gott, God
greifen,* to grasp, reach
groß, big, large, great
der Grund, basis, reason, ground
gut, good; (*adv.*) well

haben,* to have
halb, half
halten,* *v.t.* to hold, keep
       v.i. to stop, halt
die Hand, hand
Haupt . . . (*prefix*) main, chief
das Haus, house
her[s], here, to this place
herrschen, to prevail, rule

Most Frequent Words

das Herz, heart
hier, here, at this place
hin[s], there, to that place
hoch (hoh . . .), high

ich, I
ihr, 1. *possess.* her, its, their
   2. *pron.* (to) her, it
immer, always, ever
in, in, into

ja, yes, after all, in fact
das Jahr, year
jeder, each, every(one)
jener, that, *plur.* those
jetzt, now

kein, no, not a(ny)
kennen,° to know
klein, small, little
kommen,° to come, get (to)
können,° to be able

das Land, land, country
lang, long
lassen,° to let, allow, leave
leben, to live
legen, to lay, put, place
letzt, last
liegen,° to lie, be situated

machen, to make, do
das Mal, time (*as in:* dreimal, *three times*)
man, one, we, a person
der Mann, man
mehr, more
mein, my
Mensch, person, human being

messen,° to measure
mit[s], with
mögen,° to like; may
möglich, possible
müssen,° to have to

nach[s], after; according to, to, toward
nähern, to approach
der Name, name
die Natur, nature
nehmen,° to take
nennen,° to name, mention
neu, new
nicht, not
noch, yet, still, in addition
nun, now
nur, only

oder, or
ohne, without

der Rat, advice, council, councilor
das Recht, right, (field of) law
reden, to talk
reich, rich

die Sache, matter, thing
sagen, to say, tell
sammeln, to collect
der Satz, sentence; set; theorem, law
scheiden,° to separate
scheinen,° to seem; shine
schlagen, to beat, strike
schließen,° to close, conclude
schon, already
schreiben,° to write

sehen,° to see
sehr, very (much)
sein, his, its, one's
sein,° to be
Seite, side, page
selber *or* selbst, itself, oneself, themselves, etc.
setzen, to put, set
sich, himself, herself, itself, oneself, themselves
sicher, sure
sie, she, her, it; they, them
der Sinn, sense, meaning
sitzen,° to sit, be located
so, so, thus, in this way
solch, such
sollen,° to be supposed to
sonder . . . , (*prefix*) special
sprechen,° to speak
der Staat, state (*polit.*)
stehen,° stand
stellen, to put, place
das Steuer, rudder, controls
die Steuer, tax
suchen, to search, seek

der Tag, day
der Teil, part
tragen,° to carry, bear, wear
treten,° to step
tun,° to do, make; act; put

über^s, over, above, about
um^s, about, around
und, and

unser, our
unter^s, under; among

viel, much, *plur.* many
voll^s, full
von, of, from; by
vor^s, before; ago

wahr, true
was, what
weil, because
die Weise, manner, way
weit, far, wide
welch, which, what
wenig, little, not much; *plur.* few
wenn, if, when
werden,° to become
wie, how, as, like, such as
wieder^s, again; back
wir, we
wissen,° to know
wo, where
wohl, probably, well
wollen,° to want
das Wort, word

zahlen, to pay
die Zeit, time
ziehen,° *v.t.* to pull, draw
v.i. to move
zu^s, to, toward; too (= excessively)
zurück^s, back
zusammen^s, together
zwei, two

APPENDIX 6

Principal Parts of Simple Stem-Changing Verbs

Verbs with prefixes are listed only if the unprefixed form is rare or does not exist. The fourth principal part (present singular) is given only if a stem-change is involved.

INFINITIVE	PAST	PAST PARTICIPLE	PRESENT SINGULAR
backen (to bake, fry)	buk *or* backte	gebacken	bäckt *or* backt
befehlen (to command)	befahl	befohlen	befiehlt
beginnen (to begin)	begann	begonnen	
beißen (to bite)	biß	gebissen	
bergen (to conceal)	barg	geborgen	birgt
bersten (to burst)	barst	ist geborsten	birst
biegen (to bend)	bog	gebogen[1]	
bieten (to offer)	bot	geboten	
binden (to bind, tie)	band	gebunden	
bitten (to request)	bat	gebeten	
blasen (to blow)	blies	geblasen	bläst
bleiben (to remain)	blieb	ist geblieben	
braten (to roast)	briet	gebraten	brät
brechen (to break)	brach	gebrochen[1]	bricht

INFINITIVE	PAST	PAST PARTICIPLE	PRESENT SINGULAR
brennen (to burn)	brannte	gebrannt	
bringen (to bring)	brachte	gebracht	
denken (to think)	dachte	gedacht	
dreschen (to thresh)	drosch	gedroschen	drischt
dringen (to penetrate)	drang	ist gedrungen	
dürfen (to be permitted)	durfte	dürfen *or* gedurft	darf
empfehlen (to recommend)	empfahl	empfohlen	empfiehlt
erschrecken[2] (to be frightened)	erschrak	ist erschrocken	erschrickt
essen (to eat)	aß	gegessen	ißt
fahren (to travel)	fuhr	ist gefahren	fährt
fallen (to fall)	fiel	ist gefallen	fällt
fangen (to catch)	fing	gefangen	fängt
finden (to find)	fand	gefunden	
flechten (to braid)	flocht	geflochten	flicht
fliegen (to fly)	flog	ist geflogen	
fliehen (to flee)	floh	ist geflohen	
fließen (to flow)	floß	ist geflossen	
fressen (to corrode)	fraß	gefressen	frißt
frieren (to freeze)	fror	gefroren[3]	
gären (to ferment)	gor	gegoren	
gebären (to give birth)	gebar	geboren[4]	gebiert
geben (to give)	gab	gegeben	gibt
gedeihen (to thrive)	gedieh	ist gediehen	
gehen (to go)	ging	ist gegangen	
gelingen (to be possible)	gelang	ist gelungen	
gelten (to be valid)	galt	gegolten	gilt
genesen (to recover)	genas	ist genesen	
genießen (to enjoy)	genoß	genossen	
geschehen (to happen)	geschah	ist geschehen	geschieht
gewinnen (to win)	gewann	gewonnen	
gießen (to pour)	goß	gegossen	
gleichen (to resemble)	glich	geglichen	
gleiten (to glide)	glitt	ist geglitten	
glimmen (to glimmer)	glomm	geglommen	
graben (to dig)	grub	gegraben	gräbt

Principal Parts of Verbs

INFINITIVE	PAST	PAST PARTICIPLE	PRESENT SINGULAR
greifen (to grasp)	griff	gegriffen	
haben (to have)	hatte	gehabt	hat
halten (to hold)	hielt	gehalten	hält
hangen or hängen[5] (to hang)	hing	gehangen[6]	hängt
hauen (to beat, chop)	hieb or haute	gehauen or gehaut	
heben (to raise)	hob	gehoben	
heißen (to mean; be named)	hieß	geheißen	
helfen (to help)	half	geholfen	hilft
kennen (to know)	kannte	gekannt	
klimmen (to climb)	klomm	ist geklommen	
klingen (to sound)	klang	geklungen	
kneifen (to pinch)	kniff	gekniffen	
kommen (to come)	kam	ist gekommen	
können (to be able)	konnte	können or gekonnt	kann
kriechen (to crawl)	kroch	ist gekrochen	
laden (to load)	lud or ladete	geladen	lädt or ladet
lassen (to let, leave)	ließ	(ge)lassen	läßt
laufen (to run)	lief	ist gelaufen	läuft
leiden (to suffer)	litt	gelitten	
leihen (to lend)	lieh	geliehen	
lesen (to read)	las	gelesen	liest
liegen (to lie, be situated)	lag	gelegen[6]	
lügen (to lie, prevaricate)	log	gelogen	
meiden (to avoid)	mied	gemieden	
messen (to measure)	maß	gemessen	mißt
mögen (to like, may)	mochte	mögen or gemocht	mag
müssen (to have to)	mußte	müssen or gemußt	muß
nehmen (to take)	nahm	genommen	nimmt
nennen (to name)	nannte	genannt	
pfeifen (to whistle)	pfiff	gepfiffen	
preisen (to praise)	pries	gepriesen	

INFINITIVE	PAST	PAST PARTICIPLE	PRESENT SINGULAR
quellen (to well up)	quoll	ist gequollen	quillt
raten (to advise; guess)	riet	geraten	rät *or* ratet
reiben (to rub)	rieb	gerieben	
reißen (to tear)	riß	gerissen[1]	
reiten (to ride)[7]	ritt	ist geritten	
rennen (to run)	rannte	ist gerannt	
riechen (to smell)	roch	gerochen	
ringen (to wrestle, wring)	rang	gerungen	
rinnen (to flow, run)	rann	ist geronnen	
rufen (to call)	rief	gerufen	
saugen (to suck)	sog	gesogen	
schaffen (to create)[8]	schuf	geschaffen	
scheiden (to separate)	schied	geschieden	
scheinen (to seem, shine)	schien	geschienen	
schieben (to push)	schob	geschoben	
schießen (to shoot)	schoß	geschossen[1]	
schlafen (to sleep)	schlief	geschlafen	schläft
schlagen (to beat)	schlug	geschlagen	schlägt
schleichen (to sneak)	schlich	ist geschlichen	
schliefen (to slip)	schloff	ist geschloffen	
schließen (to close)	schloß	geschlossen	
schlingen (to twine)	schlang	geschlungen	
schmelzen[2] (to melt)	schmolz	ist geschmolzen	schmilzt
schneiden (to cut)	schnitt	geschnitten	
schreiben (to write)	schrieb	geschrieben	
schreien (to scream)	schrie	geschrieen	
schreiten (to stride)	schritt	ist geschritten	
schweigen (to be silent)	schwieg	geschwiegen	
schwimmen (to swim, float)	schwamm	ist geschwommen	
schwinden (to vanish)	schwand	ist geschwunden	
schwingen (to swing)	schwang	geschwungen	
schwören (to swear)	schwor	geschworen	
sein (to be)	war	ist gewesen	ist
senden (to send)	sandte *or* sendete	gesandt *or* gesendet	

INFINITIVE	PAST	PAST PARTICIPLE	PRESENT SINGULAR
sieden (to boil)	sott *or* siedete	gesotten *or* gesiedet	
sinken (to sink, *v.i.*)	sank	ist gesunken	
sinnen (to ponder)	sann	gesonnen	
sitzen (to sit)	saß	gesessen[6]	
sollen (to be supposed to)	sollte	sollen *or* gesollt	soll
speien (to spew)	spie	gespieen	
spinnen (to spin)	spann	gesponnen	
sprechen (to speak)	sprach	gesprochen	spricht
sprießen (to sprout)	sproß	ist gesprossen	
springen (to leap; crack)	sprang	ist gesprungen	
stechen (to prick)	stach	gestochen	sticht
stehen (to stand)	stand	gestanden[6]	
stehlen (to steal)	stahl	gestohlen	stiehlt
steigen (to climb)	stieg	ist gestiegen	
sterben (to die)	starb	ist gestorben	stirbt
stieben (to scatter)	stob	ist gestoben	
stinken (to stink)	stank	gestunken	
stoßen (to push)	stieß	gestoßen[1]	stößt
streichen (to stroke, paint)	strich	gestrichen	
streiten (to dispute)	stritt	gestritten	
tragen (to carry)	trug	getragen	trägt
treffen (to meet)	traf	getroffen	trifft
treiben (to drive)	trieb	getrieben[1]	
treten (to step)	trat	ist getreten	tritt
triefen (to drip)	troff	getroffen	
trinken (to drink)	trank	getrunken	
trügen (to deceive)	trog	getrogen	
tun (to do, act)	tat	getan	tut
verderben (to spoil)	verdarb	verdorben[1]	verdirbt
vergessen (to forget)	vergaß	vergessen	vergißt
verlieren (to lose)	verlor	verloren	
wachsen (to grow)	wuchs	ist gewachsen	wächst
wägen (to weigh)	wog	gewogen	
waschen (to wash)	wusch	gewaschen	wäscht
weben (to weave)	wob	gewoben	

INFINITIVE	PAST	PAST PARTICIPLE	PRESENT SINGULAR
weichen (to yield)	wich	ist gewichen	
wenden (to turn)	wandte *or* wendete	gewandt *or* gewendet	
werben (to woo)	warb	geworben	wirbt
werden (to become)	wurde	ist geworden[9]	wird
werfen (to throw)	warf	geworfen	wirft
wiegen (to weigh)	wog	gewogen	
winden (to wind)	wand	gewunden	
wissen (to know)	wußte	gewußt	weiß
wollen (to want)	wollte	wollen *or* gewollt	will
ziehen (to pull; move)	zog	gezogen[1]	
zwingen (to force)	zwang	gezwungen	

[1] As v.i. often with **ist**.
[2] As v.t. this verb is non-stem-changing.
[3] In some meanings also with **ist**.
[4] **ist geboren** or **wurde geboren** was born.
[5] As v.t. often non-stem-changing.
[6] In some regions with **ist**.
[7] On horseback.
[8] In other meanings this verb is non-stem-changing.
[9] In passive perfects this form is changed to **worden**.

APPENDIX **7**

Alphabetical List of Verb Forms

The following is an alphabetical list of all the forms (principal parts) of the stem-changing verbs included among the first 750 words in the Morgan *Frequency Word Book*. The forms listed are: the infinitive (which is also the plural of the present tense for all verbs except **sein,** to be); the singular of the past tense; the past participle; the third person singular of the present tense whenever it involves a stem change (otherwise this form can be traced to the infinitive by changing the final **-t** or **-et** to **-en**: findet—finden). Additional forms are included for the highly irregular verb **sein,** and a few irregular older forms of the short conditional (subjunctive) (**stünde** for **stände**) are also listed (otherwise short conditionals are to be reduced to the simple past by removing the umlaut and/or a final **-e,** unless this **-e** is part of a **-te** ending: **wäre—war, müßte—mußte**). Only unprefixed verbs are included, except in cases like **beginnen, gewinnen, verlieren,** etc., whose stems never occur without the prefix. Additional prefixed verbs that come within the first 750 words of the frequency list have been omitted here because they are based on unprefixed verbs already included (**erfahren—fahren, anfangen—fangen,** etc.). If a verb forms its perfects with the auxiliary **sein,** its past participle is listed here with the verb form **ist** before it in parentheses: **(ist) gegangen.**

aß, past of **essen,** to eat

band, past of **binden,** to bind
barg, past of **bergen,** to conceal
bat, past of **bitten,** to request
befahl, past of **befehlen,** to command
befehlen, to command
befiehlt, pres. sing. of **befehlen,** to command
befohlen, p. p. of **befehlen,** to command
begann, past of **beginnen,** to begin
beginnen, to begin
begonnen, p. p. of **beginnen,** to begin
bergen, to conceal, shelter
bieten, to offer
bin (am), 1st pers. pres. sing. of **sein,** to be
binden, to bind, tie
birgt, pres. sing. of **bergen,** to conceal
bitten, to request, ask
bleiben, to stay, remain
blieb, past of **bleiben,** to remain
bot, past of **bieten,** to offer
brach, past of **brechen,** to break
brachte, past of **bringen,** to bring
brechen, to break
bricht, pres. sing. of **brechen,** to break
bringen, to bring, put, take

dachte, past of **denken,** to think
darf, pres. sing. of **dürfen,** to be permitted

denken, to think
drang, past of **dringen,** to penetrate
dringen, to penetrate, press forward
dürfen, to be permitted (also p. p.)
durfte, past of **dürfen,** to be permitted

essen, to eat

fahren, to travel, ride, drive, go
fährt, pres. sing. of **fahren,** to travel
fallen, to fall
fällt, pres. sing. of **fallen,** to fall
fand, past of **finden,** to find
fangen, to catch
fängt, pres. sing. of **fangen,** to catch
fiel, past of **fallen,** to fall
finden, to find
fing, past of **fangen,** to catch
fliegen, to fly
fliehen, to flee
fließen, to flow
flog, past of **fliegen,** to fly
floh, past **fliehen,** to flee
floß, past of **fließen,** to flow
fraß, past of **fressen,** to corrode
fressen, to corrode, eat away
frißt, pres. sing. of **fressen,** to corrode
fuhr, past of **fahren,** to travel

gab, past of **geben,** to give

Alphabetical List of Verb Forms

galt, past of **gelten**, to be valid
geben, to give
gebeten, p. p. of **bitten**, to request
(ist) geblieben, p. p. of **bleiben**, to remain
geborgen, p. p. of **bergen**, to conceal
geboten, p. p. of **bieten**, to offer
gebracht, p. p. of **bringen**, to bring
gebrochen (also v.i.: ist gebrochen), p. p. of **brechen**, to break
gebunden, p. p. of **binden**, to bind
gedacht, p. p. of **denken**, to think
(ist) gedrungen, p. p. of **dringen**, to penetrate
(ist) gefahren, p. p. of **fahren**, to travel
(ist) gefallen, p. p. of **fallen**, to fall
gefangen, p. p. of **fangen**, to catch
(ist) geflogen, p. p. of **fliegen**, to fly
(ist) geflohen, p. p. of **fliehen**, to flee
(ist) geflossen, p. p. of **fließen**, to flow
gefressen, p. p. of **fressen**, to corrode
gefunden, p. p. of **finden**, to find
(ist) gegangen, p. p. of **gehen**, to go
gegeben, p. p. of **geben**, to give
gegessen, p. p. of **essen**, to eat
gegolten, p. p. of **gelten**, to be valid
gegraben, p. p. of **graben**, to dig
gegriffen, p. p. of **greifen**, to grasp
gehabt, p. p. of **haben**, to have
gehalten, p. p. of **halten**, to hold, etc.
gehangen, p. p. of **hängen**, to hang
geheißen, p. p. of **heißen**, to mean
gehen, to go
gehoben, p. p. of **heben**, to lift
geholfen, p. p. of **helfen**, to help
gekannt, p. p. of **kennen**, to know
(ist) gekommen, p. p. of **kommen**, to come
gelang, past of **gelingen**, to be possible
gelassen, p. p. of **lassen**, to let
(ist) gelaufen, p. p. of **laufen**, to run
gelegen, p. p. of **liegen**, to lie
gelesen, p. p. of **lesen**, to read
gelingen, to be possible, to be successful
gelingt, pres. sing. of **gelingen**, to be possible
gelitten, p. p. of **leiden**, to suffer
gelten, to be valid, hold true, to be considered
(ist) gelungen, p. p. of **gelingen**, to be possible
gemessen, p. p. of **messen**, to measure
genannt, p. p. of **nennen**, to mention
genießen, to enjoy
genießt, pres. sing. of **genießen**, to enjoy
genommen, p. p. of **nehmen**, to take
genoß, past of **genießen**, to enjoy

genossen, p. p. of **genießen,** to enjoy

gerissen, (also v.i.: ist gerissen), p. p. of **reißen,** to tear

gerufen, p. p. of **rufen,** to call

gesandt, p. p. of **senden,** to send

geschah, past of **geschehen,** to happen

geschehen, 1. (inf.) to happen, be done 2. (ist) **geschehen,** p. p.

geschieden, p. p. of **scheiden,** to separate

geschieht, pres. sing. of **geschehen,** to happen

geschienen, p. p. of **scheinen,** to seem

geschlafen, p. p. of **schlafen,** to sleep

geschlagen, p. p. of **schlagen,** to beat

geschlossen, p. p. of **schließen,** to conclude

geschnitten, p. p. of **schneiden,** to cut

geschossen (also v.i.: ist geschossen), p. p. of **schießen,** to shoot

geschrieben, p. p. of **schreiben,** to write

(ist) geschritten, p. p. of **schreiten,** to step

geschwiegen, p. p. of **schweigen,** to be silent

gesehen, p. p. of **sehen,** to see

gesendet, p. p. of **senden,** to send

gesessen, p. p. of **sitzen,** to sit

gesprochen, p. p. of **sprechen,** to speak

(ist) gesprungen, p. p. of **springen,** to leap

gestanden, p. p. of **stehen,** to stand

(ist) gestiegen, p. p. of **steigen,** to climb

(ist) gestorben, p. p. of **sterben,** to die

gestritten, p. p. of **streiten,** to dispute

(ist) gesunken, p. p. of **sinken,** to sink

getragen, p. p. of **tragen,** to carry

(ist) getreten, p. p. of **treten,** to step

getrieben, p. p. of **treiben,** to drive

getroffen, p. p. of **treffen,** to meet

getrunken, p. p. of **trinken,** to drink

(ist) gewachsen, p. p. of **wachsen,** to grow

gewandt, p. p. of **wenden,** to turn

gewann, past of **gewinnen,** to win

gewendet, p. p. of **wenden,** to turn

(ist) gewesen, (been) p. p. of **sein,** to be

(ist) gewichen, p. p. of **weichen,** to yield

gewiesen, p. p. of **weisen,** to show

gewinnen, to win, gain, obtain

gewinnt, pres. sing. of **gewinnen,** to win

gewogen, p. p. of **wiegen,** to weigh

gewonnen, p. p. of **gewinnen,** to win
(ist) geworden (become), p. p. of **werden**
geworfen, p. p. of **werfen,** to throw
gewußt, p. p. of **wissen,** to know
gezogen (also v.i.: ist gezogen), p. p. of **ziehen,** to pull, etc.
gezwungen, p. p. of **zwingen,** to force
gibt, pres. sing. of **geben,** to give
gilt, pres. sing. of **gelten,** to be valid
ging, past of **gehen,** to go
graben, to dig
gräbt, pres. sing. of **graben,** to dig
greifen, to grasp, reach
griff, past of **greifen,** to grasp
grub, past of **graben,** to dig

haben, to have
half, past of **helfen,** to help
hält, pres. sing. of **halten,** to hold
halten, 1. v.t., to hold, keep
    2. v.i., to stop, come to a halt
hangen, hängen, to hang
hängt, pres. sing. of **hängen,** to hang
hat, pres. sing. of **haben,** to have
heben, to lift, raise
heißen, to mean, to be named
helfen, to help, aid
hielt, past of **halten,** to hold, etc.
hieß, past of **heißen,** to mean, etc.
hilft, pres. sing. of **helfen,** to help
hing, past of **hängen,** to hang

hob, past of **heben,** to lift
ißt, pres. sing. of **essen,** to eat
ist (is), 3d pers. pres. sing. of **sein,** to be

kam, past of **kommen,** to come
kann (can), pres. sing. of **können**
kannte, past of **kennen,** to know
kennen, to know, be acquainted with
kennte, short cond'l. of **kennen,** to know
kommen, to come, get (to)
können, to be able (also p. p.)
konnte (could, was able), past of **können**

lag, past of **liegen,** to lie
las, past of **lesen,** to read
lassen, to let, allow, to leave (also often p. p.)
läßt, pres. sing. of **lassen,** to let
laufen, to run
läuft, pres. sing. of **laufen,** to run
leiden, to suffer, tolerate
lesen, to read
lief, past of **laufen,** to run
liegen, to lie, be situated
ließ, past of **lassen,** to let
liest, pres. sing. of **lesen,** to read
litt, past of **leiden,** to suffer

mag (may; like), pres. sing. of **mögen**
maß, past of **messen,** to measure
messen, to measure
mißt, pres. sing. of **messen,** to measure
mochte (might; liked), past of **mögen**

möchte (would like), short cond'l. of **mögen**
mögen, to like, (also p. p.) (pres. and past often: may, might)
muß (must, has to) pres. sing. of **müssen**, to have to
müssen, to have to (also p. p.)
mußte, past of **müssen**, to have to

nahm, past of **nehmen**, to take
nannte, past of **nennen**, to name
nehmen, to take
nennen, to name, call, mention
nennte, short cond'l. of **nennen**, to name
nimmt, pres. sing. of **nehmen**, to take

reißen, to tear, rip; draw (= depict)
rief, past of **rufen**, to call
riß, past of **reißen**, to tear
rufen, to call, cry (out)

sah, past of **sehen**, to see
sandte, past of **senden**, to send
sank, past of **sinken**, to sink
saß, past of **sitzen**, to sit
scheiden, to separate, divorce
scheinen, to seem, to shine
schied, past of **scheiden**, to separate
schien, past of **scheinen**, to seem
schießen, to shoot, v.i. oft., to dart
schlafen, to sleep
schläft, pres. sing. of **schlafen**, to sleep
schlagen, to beat, strike, fold
schlägt, pres. sing. of **schlagen**, to beat

schlief, past of **schlafen**, to sleep
schließen, to close, conclude
schloß, past of **schließen**, to conclude
schlug, past of **schlagen**, to beat
schneiden, to cut, intersect
schnitt, past of **schneiden**, to cut
schreiten, to step, stride
schritt, past of **schreiten**, to step
schoß, past of **schießen**, to shoot
schreiben, to write
schrieb, past of **schreiben**, to write
schweigen, to be silent, say nothing
schwieg, past of **schweigen**, to be silent
sehen, to see; (esp. with prefix) to look
sei, exhortation and quoting form of **sein**, to be
sein, to be
senden, to send
sendete, past and esp. short cond'l. of **senden**, to send
siehe, imperative of **sehen**, to see
sieht, pres. sing. of **sehen**, to see
sind (are), pres. plur. of **sein** to be
sinken, to sink, go down, drop
sitzen, to sit, be located
soll, pres. sing. of **sollen**, to be supposed to
sollen, to be supposed to (also p. p.)
sollte, 1. past of **sollen**, to be supposed to
2. (should, ought to) short cond'l. of **sollen**
sprach, past of **sprechen**, to speak

sprang, past of **springen,** to leap
sprechen, to speak
spricht, pres. sing. of **sprechen,** to speak
springen, to leap, jump, to crack
stand, past of **stehen,** to stand
starb, past of **sterben,** to die
stehen, to stand
steigen, to rise, climb
sterben, to die
stieg, past of **steigen,** to climb
stirbt, pres. sing. of **sterben,** to die
streiten, to dispute, quarrel, fight
stritt, past of **streiten,** to dispute
stünde, older form of short cond'l. of **stehen,** to stand
stürbe, older form of short cond'l. of **sterben,** to die

traf, past of **treffen,** to meet
tragen, to carry, bear, wear
trägt, pres. sing. of **tragen,** to carry
trank, past of **trinken,** to drink
trat, past of **treten,** to step
treffen, to meet, to strike
treiben, to drive, impel
treten, to step, tread, enter
trieb, past of **treiben,** to drive
trifft, pres. sing. of **treffen,** to meet
trinken, to drink
tritt, pres. sing. of **treten,** to step
trug, past of **tragen,** to carry

verlieren, to lose

verlor, verloren, past and p. p. of **verlieren,** to lose

wachsen,[*] to grow, wax, increase
wächst, pres. of **wachsen,** to grow
wandte, past of wenden, to turn
war (was), waren (were), past of **sein,** to be
warf, past of **werfen,** to throw
weichen, to yield, give way
weisen, to show, point (to)
weiß, pres. sing. of **wissen,** to know
wenden, to turn
wendete, past and esp. short cond'l. of **wenden,** to turn
werden, to become, get, grow (in pres. tense, aux. for future: will); (in all tenses, aux. for passive: to be, get)
werfen, to throw
wich, past of **weichen,** to yield
wiegen, to weigh
wies, past of **weisen,** to show
will, pres. sing. of **wollen,** to want
wird, pres. sing. of **werden,** to become, etc.
wirft, pres. sing. of **werfen,** to throw
wissen, to know
wog, past of **wiegen,** to weigh
wollen, to want (also p. p.)
wollte, past and short cond'l. of **wollen,** to want
worden, (been) p. p. of **werden,** in passive perfect tenses

---

[*] Do not confuse with **waschen,** to wash.

wuchs, past of **wachsen,** to grow
wurde, past of **werden,** to become, etc.
wußte, past of **wissen,** to know

ziehen, 1. v.t., to pull, draw
    2. v.i., to move
zog, past of **ziehen,** to pull, etc.
zwang, past of **zwingen,** to force
zwingen, to force